QUANTUM PHYSICS

A Simple Guide to Understanding the Principles of Physics

(All the Major Ideas of Quantum Mechanics From Quanta to Entanglement in Simple Language)

John Hernandez

Published By John Hernandez

John Hernandez

All Rights Reserved

Quantum Physics: A Simple Guide to Understanding the Principles of Physics (All the Major Ideas of Quantum Mechanics From Quanta to Entanglement in Simple Language)

ISBN 978-1-77485-414-3

Legal & Disclaimer

The information contained in this book is not designed to replace or take the place of any form of medicine or professional medical advice. The information in this book has been provided for educational and entertainment purposes only.

The information contained in this book has been compiled from sources deemed reliable, and it is accurate to the best of the Author's knowledge; however, the Author cannot guarantee its accuracy and validity and cannot be held liable for any errors or omissions. Changes are periodically made to this book. You must consult your doctor or get professional medical advice before using any of the suggested remedies, techniques, or information in this book.

Upon using the information contained in this book, you agree to hold harmless the Author from and against any damages, costs, and expenses, including any legal fees potentially resulting from the application of any of the information provided by this guide. This disclaimer applies to any damages or injury caused by the use and application, whether directly or indirectly, of any advice or information presented, whether for breach of contract, tort, negligence, personal injury, criminal intent, or under any other cause of action.

You agree to accept all risks of using the information presented inside this book. You need to consult a professional medical practitioner in order to ensure you are both able and healthy enough to participate in this program.

TABLE OF CONTENTS

Introduction

Quantum Physics, as with every other field of study is a quest to define what is real. While quantum physics is an art, its enthralling results have blurred the lines between the classical sciences and their ancient rivals in religion, philosophy and mysticism.

In the end, the quantum investigation hasn't proved truth, but instead redefined what it means to be. It's like trying to tune the car, and then somewhere during the process we found we could see that it was not an automobile, but rather it was a flubbermobile. As Someone said, "Not only is the Universe more mysterious than we imagine but it's even stranger than we think." This is certainly a true assertion when trying to grasp quantum Physics.

Yet the basic ideas of quantum physics can make sense as long as you keep them at an understanding level. Take a look at the following three fundamental concepts:

1. The fundamental tenet that is the basis of quantum physics there is a vast infinite number of possibilities at the most fundamental the level of matter. It is what is known as"the "unified field." Matter and form all stem from this mysterious, nebulous web of potentials.

2. Studies with subatomic particles have proven that time and space are not relevant at this impossible scale of matter. The particles can not only be seen as particles or waves but they also can be present in two locations (called "superposition"). Quantum science has managed to provide evidence that shows not only do past experiences influence the present (a discovery that we've been taught to expect) however, future events may also create changes to the present real-world reality.

3. Connected in related to "construction that the brain makes that is designed to produce order and form in the midst of chaos in the universe." From 1927

onwards scientists began to understand the significance and impact of the observer's perspective in any given reality. The Heisenberg Uncertainty Principle resulted from the discovery that, not only is a particle (reality) change as it is observed, but that the moment an observer changes the way they view or perceive a certain real world, then that reality alters in tandem with it.

Therefore, there is no any absolute truth, or objective reality. It's everything subjective.

What would that mean If that were the case? A few people might raise their hands and declare, "Let anarchy begin," or "This notion will lead to an environment of egomaniacs" or "How do we approach universal reality such as Nature, Justice, and Truth?"

In response, one could say, "Chaos reigns without us giving him permission" or "Take one look around the world; haven't you already created a world of people who are egomaniacs?" Or "The world is always fighting because of the inexplicably individual interpretations of religion as well as Justice and Reality."

If you stop and think about the question with an open mind it becomes clear how our experiences of the world are subjective. The most obvious proof is the fact that in the event that twenty-five people had attended the same class there would be various different experiences of the same event.

In spite of this the fact that anything or anything is subjective, is controversial. So it's not a surprise that nearly eighty years after the very first "quantum" discoveries and we're still trapped in an idea that is based on one simple and absolutely true fact. It is unfortunate that there is no place for creativity, awareness or faith in this universally accepted "clockwork the universe." It's a ideal place to relax for those who do not value (or more importantly, don't trust) the power of imagination, morality, belief, or the unity of.

As a result of the advances in quantum Physics, we'll be required ways of thinking differently. The new 21st century offers the possibility to build a new heaven an entirely new planet, and, to use the word that has been utilized to describe a new way of understanding the universe. There is a whole new world waiting for us. If you're interested in the quantum physics world I strongly recommend you read each theory and law discussed in this book. We will travel to the quantum physics realm.

Chapter 1: What Exactly Is Quantum Physics?

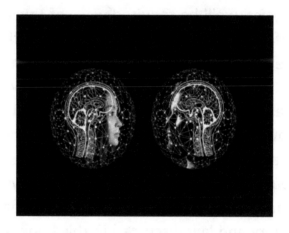

Quantum Mechanics, commonly referred to as Quantum Physics, is the interaction between matter and energy. The word "quantum" refers to Latin meaning "how much." The mechanics of this term refers to a the quantum theory attributes to specific physical phenomena in tiny amounts as a way of measuring. Quantum expressions are typically looked at and studied on the sub-atomic level using sub-atomic particles.

Subatomic particles are extremely small. If an atom was as big as a room it would be the same size like the drop of gum inside the kitchen cabinet of the building. There are several things that happened before the concept of Quantum Mechanics took root. In 1838, following discovering cathode radiation, Gustav Kirchoff, in 1850, released an opinion piece regarding"black bodies" and the "black bodies radiation" issue. Then, in 1877 Ludwig Boltzmann proposed that the state of energy in the physical system could be asymmetrical.

In the year 1900, Max Planck came up with the idea that energy is radiated and absorbent, and later the formula he devised will be referred to by the name of "Planck 's Constant for Action."

Planck is also considered to be Quantum Mechanics' patriarch. After the publication of his theory and accepted by other scientists, they took note of it and there were a number of theories to develop prior to Quantum Mechanics was theorized and was studied across the globe.

It's due to Quantum Mechanics that we're on the brink of antigravity and super conductors. MRI equipment in hospital settings, and we're also able to envision that time travel is feasible.

This is all so fascinating It's true, but these are the things that researchers working in the field of quantum mechanics will be able to share with you. The most difficult aspect for the majority of us to comprehend is the relationship between sub-atomic particle and The Law of Attraction.

The Quantum Mechanics test, it was observed that sub-atomic particles are moving in a particular direction. Every other force moves the fundamental elements of physical matter throughout the universe.

After a few double-blind studies that used 'Sub-Atomic particles as the subject matter, researchers realized that they can change into waveforms from particles and back. They could even be removed from our dimensions and then pop into it again. We also observed that sub-atomic particles dependent on their intent they have transformed into waveforms instead of particles. We discovered that we couldn't eliminate the equation during testing of the particles. We've influenced the particles because we were worried about the outcome. There's much of more.

It's becoming extremely complex. Einstein was unsure till the time he passed away. Understanding the duality of wave and particle isn't something we can easily wrap our minds around.

One of the concepts that have emerged out of Quantum Physics is that we affect the very structure of life through our thoughts about it. Our minds contain a thought that is released and back to us with what we're thinking about. This is the Law of Attraction.

A clock that is moving moves at a slow pace as its speed increases and then it starts to run in the light speed except if you are traveling at the same speed as the clock and in that case, it behaves just as it normally does. When speeds of objects rises it's apparent size decreases, while its weight rises up until it reaches the point where light speed has been attained, and then it disappears and its weight becomes inexhaustible. It is fortunate that nothing is discernible until you get close to speed at which light travels. Similar to that quantum physics has proven that even though the universe appears to be logical and predictable in the subatomic realm (dealing with the various elements that is the basis of the atoms) however, the reality is quite different. For example:

* You can show that subatomic entities are both matter and energy. It's all dependent on the tests you perform. It is not possible to be an objective observer. What you see is bound to preconceptions of your personal beliefs.

You cannot predict what a subatomic object is going to do, but you just know the things it's going to accomplish. Therefore, the universe can no longer be considered an inanimate object that follows certain logical principles.

* Entities that share the same source are connected, even though they're not, meaning that whatever happens to one is mirrored within the second. In this scenario, the connections between entities are much greater than individual identities, and even the entire universe may be connected.

In an effort to comprehend it all, the theoretical physicist David Bohm, suggests that reality is not only composed of matter and energy instead, it is composed made up of material (explained order) and meaning (implicit order) along with the concept of meaning (super-implicit order). Each of them contains the other two- matter that is composed of meaning and energy, thirds energy and meaning, as well as the perception of energy and matter. The truth, therefore, is not two- dimensional in the sense of an image, but is multidimensional in the sense of an hologram, with every element being able to represent the entire and the entire to be understood.

Certain people make use of quantum physics for the purpose of "prove" the truthfulness of what magicians and mystics were saying for centuries- but do they really have to be able to prove that? Furthermore, much of quantum Physics is conceptual. What's been accomplished is to close the gap that exists between science and religious. Buddhism is usually regarded as a path to spiritual awakening for those with an intellect. This is also true for contemporary Physics. We will continue to flounder with our usual way of life and, who is to say? We'll all end at the same spot.

Quantum science is at the heart of the way atoms function and the reason why chemistry and biology function as they do. I, you, and the gatepost, in some way we're all dancing to the quantum beat. If you're interested in understanding the way electrons move through an electronic chip, the way light particles change into electrical currents in solar panels or amplify their energy in a laser or even the way that the sun's heat is generated in the sun, you'll have to employ quantum Physics.

The complexity, and for physicists the fun starts here. The first thing to note is that there isn't a single quantum theory.

Quantum mechanics is the basic mathematical structure that is the basis of everything that was first developed around 1920 in the 1920s by Niels Bohr Werner Heisenberg, Erwin Schrodinger and many others. It explains simple things such as how the momentum or position of one particle or a group of a few particles shifts in time.

To fully comprehend how things function on a daily basis, quantum mechanics has to be paired with other aspects of physics, including Albert Einstein's special theory, which explains the way objects move extremely quicklyand to create what's called quantum field theories.

Three quantum field theories focus on one of four primary force that the universe interacts with. Electromagnetism which is the reason atoms stay together and how they are able to move. Strong nuclear power, which is the reason for the stabilization of the nucleus that lies at the center of the atom as well as weak nuclear power that explains the reasons why certain atoms are subject the radioactive decay.

Over the past 50 years or more the three theories have been compiled into a loose coalition that is referred to as"the "normal theory" of particle Physics. In spite of the fact that the model is held by sticky tape, it's the most exact depiction of the fundamental work of the subject that has been created. The ultimate highlight was achieved in 2012 when discovering the Higgs boson which is a particle that provides mass to the others fundamental particle, its existence that particle was predicted by quantum theories as long in 1964.

Quantum field theories are used to explain well to explain the results of research using particle crushers with high energy like CERN's Large Hadron Collider, where the Higgs have been discovered which is the largest in terms of size. But , you must comprehend how things function in a more abstract way - the way electrons travel travel through a solid substance to form a substance for instance, a metal, an insulation or a semiconductor for example - things can get complex.

Billions of people and billions of experiences in these crowded environments require the development of "successful field theories" which ignore some of the disturbing details. The difficulty in developing these theories is the reason that so many crucial questions about solid-state physics are unanswered. For instance the reason why, at low temperatures, specific materials can be superconductors which require electrical resistance without current and the reason we cannot make this trick work at temperatures of room temperature.

However, there's a huge quantum mystery that lies beneath the plethora of practical issues. On a fundamental level quantum physics is able to predict odd things about how the universe works. These theories do not match the way things are operating within the actual world. Quantum particles may behave as particles, residing in one place or, they may behave like waves, dispersed throughout space or spread across several locations at the same time. What they do depends on the method we choose to measure them. And prior to measuring them, they appear to lack any specific properties whatsoever, leading us to a tangled mess concerning the existence of a fundamental truth.

The fuzziness can lead to obvious contradictions, such as the Schrodinger's cat, where cats are left alive and dead at the same time due to an undetermined quantum process. However, this isn't the only thing that happens. Quantum particles also appear to interact instantly even though they're separated from each other. This phenomenon that resembles bamboo is referred to as entanglement or, as a term invented by Einstein (a famous critique of quantum theory), "spooky behavior at an infinity." Quantum powers are unfamiliar to us, however they are the basis for new technologies like quantum cryptography that is ultra-secure and secure, as well as quantum computing.

However, no one is sure the specifics of what quantum physics is about. Many people believe that we need to accept that quantum Physics describes the physical world in a way that it is unable to be a part of the wider, "classical" world. There is an even more profound, intuitive explanation available which we're yet to find.

There are a lot of things to consider in all this. As an example, there's the fourth fundamental force of nature that has not been able to define quantum theory. Gravity is still the domain in Einstein's broad Theory of Relativity, a completely non-quantum theory that does not even contain particles. The years of intensive efforts to bring gravity into the realm of quantum, and consequently provide all the fundamental physics in an "theory that is all" but have failed to produce any results.

Meanwhile the cosmological data indicate that over 95 percent of the universe is composed of dark energy and dark matter and dark energy, phenomena for which we have no reason using the standard model and also a variety of controversies such as the nature of quantum physics within the chaotic nature of the universe remain unsolved.

What is the reason that some objects can be described by the classical physics models, but others require descriptions of quantum Physics?

There are two main motives for this: coherence and size and coherence, which are briefly explained in this article. Smallness could refer to various aspects of objects, such as tiny size or low energy content. If the object is equivalent to the dimensions of one atom (about 10 meters) that means it could probably not be correctly described using classical mechanics and therefore must be described using an accurate quantum theory. It is interesting to note that this isn't necessarily an issue; objects as large as millimeters (about 20 percent or an inch) were observed during experiments that display behaviours that suggest the existence of quantum phenomena.

For instance, the tiny (or lower) or low energy level might be a reference to a small electrical current flowing through a metal wire (a superconductor) at a temperature just above the absolute zero (273 ° Celsius or 459°F). Low temperature implies low energy. It could also apply to the brief flash of light, which is the smallest fraction (say 10--21) of energy released by a bulb that produces one hundred watts for a single second. A flash of light is believed to be just one photon of light, which is the smallest, tiniest amount of energy the light of a specific hue can hold. A small amount of energy like it is

often referred to as"light quantum..'
Quantum is the plural form of quantum.'
For instance, a burst of light carrying a
huge quantity of energy can be thought to
contain several quantums. The
sluggishness of the energy contained in
light, which we'll investigate in greater
detail in the future is the reason for the
term 'quantum physics.'

In theory it is possible for a quantum
entity like photons, can be stretched over
a vast area, many kilometers for instance.
Even though a photon could be huge in
size, it will be small or very extremely low
in energy so quantum theory would be
applied to it. Another reason why an
object may require an quantum definition
is called 'quantum coherence.' Quantum
coherence is an elusive concept that is not
understood until one knows the way in
which the state for an object can be
described through quantum theory. For
the electron, quantum coherence is a part
of the concept of how it explains the many

possibilities that might exist prior to the time the electron is determined. In a sense, the common laws of logic that say, "It is located here or not," don't apply on quantum phenomena. It is instead, "Both possibilities must be thought of as a whole and not considered in isolation."

How was Quantum Physics Discovered?

The full background of how quantum physics came to be discovered is fascinating and many books tell that. However, in hindsight I think that the battle to create quantum theories in the beginning of 20th century, speaks more to the challenges that humankind faced (and continue to have) in transcending traditional physics thinking than what quantum physics' facts reveal. This book isn't focused on the scientific aspects of physics in the past. The historical details are included in this book because they help to clarify the physics discussed. In this book, I provide an outline of some of the most important historic highlights and explain the ways in which each contributed to the expanding collection of information about quantum Physics. In addition I will introduce other quantum physics concepts that are not discussed previously and, in particular the notion that quantum fields are a part of.

Scientists and philosophers of the past such as Newton asked questions regarding how light works. Are they particles or waves, or both? However, it wasn't until the year 1900 that scientifically-proven evidence was collected that began to answer this question. The story is like this When a normally black materials is heated to extreme temperatures, it releases various colors of light as if it were a metallic burner in a cooker. If the material is hot enough to be able to emit light, it will become bright enough. Light's spectrum is divided by a prism and the brightness of each hue can be measured by the use of a light detector. If these brightnesses that depend on color were compared with the predictions of the classical physics theory they were discovered to be incorrect. The German scientist Max Planck discovered that the problem with classical theories is in the logical belief that energy can transfer between hot materials and light in any amount within a continuum of energy. To find an connection between the experimental results and the theoretical model, Planck tried to change only one part of the model. He made the bold assumption that was radical for the time that the energy exchanged between the material and the light weren't continuous, but discrete; which means that they take place in stages, such as the staircase.

It was concluded by the researcher that amount of these energy levels is proportional to frequency and is in correlation by the colour of light being examined. This constant is referred to by the name of Planck constant. In scientists and physicists the new model, once mathematically formulated was in perfect agreement with the measurements made by scientists of the various brightnesses in color. Albert Einstein was motivated by the success of Planck in presenting an overall light hypothesis. Einstein thought that the light spectrum of a certain color might have only an isolated energy content that was non-continuous, like one would expect from classical Physics. He named these hidden amounts of energy as a "light quantum.' We now call them photons.

Einstein also believed that light quantums are inseparable and therefore interact in a unified way with materials that absorb light. Every photon is absorption or but it is not able to be fully taken in. He incorporated these concepts into a theoretical framework which accurately predicted the way that the atoms absorb light and emit it. His calculations proved to be, fifty years after, the scientific premise that led to the development of the laser in the year 1960. Another example of the fact that discoveries of science fundamental tend to be in the midst of the most significant technological advancements, even though the time-frame can be quite long. After Planck was pondering the rainbow-like, smooth spectrum of light that is emitted through hot materials, researchers studied the light produced by vapor or gas containing the atoms of a single element (say neon) as the current of electricity is passed through it. This is what we see in fluorescent lamps every all day. In the past it was understood that the neon atom consisted of a nucleus with 10 protons, and typically 10 neutrons that were enclosed by 10 electrons. It was also understood that electrons are "matter" since they are mass-based in contrast to photons, with zero mass. It was believed that electrons behaved as tiny planets that orbit the nucleus as if they were tiny suns. In this version of the electron, the the classical physics theory predicted that when atoms with extra energy that is retained through its electrons shed part of that energy the illumination of any hue in a continuous range could be released. But , what the experimenters discovered was that the light released comprises just a few distinct shades, not the seamless spectrum of shades. It was a huge mysterybecause the traditional physics theories did not explain this phenomenon.

To cut a long story shorter In 1925 that the problem with the original theory was the idea that electrons behaved as tiny planets. Also electrons shouldn't be considered as tiny particles or fragments of matter that travel those routes through the nucleus. Louis de Broglie -- at the time an Physics master's student in the University of Paris -- had previously speculated that electrons behave as waves, which was a non-particle-like understanding of electrons. Erwin Schrodinger was able to transform this notion into a mathematical concept which could accurately forecast all possible wavelike patterns an electron might make in a particular kind of an atom. He proved that any pattern that appears like a wave is connected to the specific frequency of the wave which, according to the concept of Planck, has an energy level that is specific to. He also found that when an electron changes from one pattern of higher energy to one with lower energy the light emits an exact frequency and consequently, a particular color. Schrodinger's equation was able accurately describe all the distinct variations in color that can be observed in light bulbs made of atomic gas.

Chapter 2: Quantum Physics - The Localization Of The Manifestation

Quantum physicists refer to electrons, which are things that have potential instead of tangible physical things. They also say that there are many potentialities, merely until you look, and that forces the universe to take a decision in a sense, about which possibilities are going to be realized and localized. Everything is essentially an endless Quantum Field of Energy an endless ocean of possibilities that are waiting to unfold!

The mind is the one who constructs and manipulating reality. Our thoughts are able to influence our world. This is the way that it works. Law of Attraction functions. We receive what we are focused on the majority of the time. The observer is simply creating reality by watching.

The Mind regardless of the form it may appear to have holds images. And any image tightly held within the mind, regardless of shape, will eventually be released.

If the mind makes thoughts or images, or even an image of something is 'one' with the Infinite Universal Consciousness and the image that is created can be seen externally into the physical world in a single time-space-event. But, to manifest the image it must not have other thoughts that conflict that could block the power of manifestation of the image stored within the mind.

Another characteristic of the quantum is that it is multi-dimensional. We can see that scientifically that the universe we live in is multidimensional even when our senses can only detecting width, length as well as height and time as dimensions. But our souls are multidimensional. Pay attention to your soul and your emotions.

The physical universe is generally composed of thoughts and energy. Many Quantum Physicists such as Einstein have proven that all physical matter is made up of energy packets which do not have a fixed location or time.

There aren't any boundaries clearly defined within this field of energy. The world is an infinite, immortal, and unbound body. The science has also proven that the mind is without boundaries. Every mind is connected to a mind-energy field. You're stronger and more powerful than you think.

Whatever you're looking for You can have it all. According to some, it will be given to you prior to your have asked. The science is slowly getting to grips with it by using quantum physics to prove that is scientifically proven. The vast knowledge of unformless substance, the possibility of quantum levels and our own inherent ability to affect the quantum field is the reason we feel of 'having everything.'

You already have the money that you can ever desires. We are slowly beginning to discover this on a bigger level, both in terms of science as well as spiritually. It's there. It may not be felt in the moment but you certainly are carrying it.

There are two kinds of aspects to be aware of and master. An easy way to grasp this is that you are able to scale Mount Everest or go Paragliding yet you may not have had the opportunity to test this aspect of your abilities. There's nothing you have to do to achieve that feat; it's already within you. It's been taught to you already. All you have to do is learn this expertise.

Quantum fields can take on endless forms, shapes , and experiences. It's likely that it's already accomplished this. The text in this book is only one of the things. The words you're reading about aren't the only things. The next thought you'll get is only one of them.

Did you think that you'll encounter these terms on these pages? The desire to discover these terms has caused them to come into your vocabulary. Actually, they've always been there. However, because of the love that you and many others have sent across the globe I've been inspired to share these thoughts with you!

There's no need to anticipate precisely how things will turn out. All you have to do is wish to do, plan, and believe that you can do it, and it'll be brought to you. Through our daily lives, we're shifting our awareness to experience the aspects of ourselves have always existed in a world which is filled with everything we could wish for. We even have something we've never imagined.

Scientists studying subatomic particles are beginning to uncover many fascinating facts about our universe. For instance, they have found that particles separated by time and space are 'invisibly connected' to one another and work in sync. They also discovered that the universe that we live in appears to be built in a manner that it can be understood by its own.

It appears that this was accomplished by cutting the 'One' total into two pieces One half of which is that is programmed to see, and the other half programmed to be observed. The one that is trained to see then has the illusion of being separated from the one who is programmed to see. It's a needless illusion. In reality all things are 'One.'

Chapter 3: Quantum Theory - An Introduction To The Mystifying Science

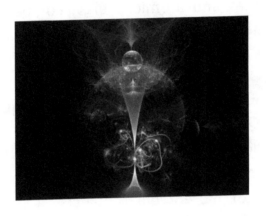

Quantum theories are the largest, most crucial exciting, intriguing, difficult and even awe-inspiring subject in science. But it's much more than just a bit bizarre. It is also one of the most inspiring concept that is being discussed in the present. It is believed that we could be wildly wrong in our beliefs about what is actually the truth.

The theory was initially called quantum mechanics because it was thought that there had to be some common principles that were involved in the work of quantum particles and atomic energy, comparable to the mechanics that govern the macro-scale subject matter of the major planets. The hypothesis aims to explain the behavior of extremely small objects, typically the magnitude of atoms , or smaller, much in similar ways to how Einstein's relativity theory defines the rules of more regular objects. It is also used in other campaigns, such as computers and TV and also helps explain the nuclear activity within and around stars.

Quantum scientists have us living in a myriad of dimensions, residing with "probability waves" and unrecognized virtual particles that flash into and out of existence however, they also declare verbally that in the future, we could travel through wormholes in the Universe to explore different cosmologies, or travel backwards through time. In simpler terms Quantum theory is the science of the leaps between one energy echelon and then to another in relation to the structure and behavior of atoms.

in 1905 Albert Einstein proposed that light was a particle not a wave. This questioned the findings of a century of research. Einstein believed that not only the energy but also the radiation itself was measured in the same manner. This is what prompted Einstein's well-known statement that "God doesn't gamble.' Einstein definitely could not accept the concept of a science that is incredibly successful because quantum mechanics could be able to give probabilities of the way that unique particles behave and don't provide certain conclusions. Because of this regardless of his many innovative methods, Einstein could never let leave the goal of pre-quantum science in being able to anticipate the universe's events just like clockwork. Quantum science isn't like Einstein considered it was, in reality, a sloppy science in fact, an extremely practical science because it acknowledges that even in the most complex of methods, science will only give rise to expectations about the reactions of different segments.

Absolutely, Quantum Theory and Albert Einstein's Theory of Relativity form the basis of contemporary physics, and nearly everyone believes that it's, in reality the theory of the invisible sphere, of small particles and huge accelerators. For the majority of people but, it's an expression for mysterious incomprehensible science. It does provide a greater range beyond the small globe and can be used to methods wherein distinct parts interact with each and also influence the other.

Quantum Physics and the Law of Attraction

It's sometimes difficult to grasp how the universe functions and how to get what you want and often, you don't find it. It is believed that the Law of Attraction and Quantum Physics combine to bring about equilibrium within the universe. It is essential to know both of them to be able to understand how the Universe functions.

First , the Law of Attraction - along with Quantum Physics - boils down to a fundamental aspect you must know to make effective utilization in this Law of Attraction. Like attracts like. It is essential to be aware of this when dealing in The Law of Attraction, so you are aware of how you are able to make use of all the laws, and know what it signifies.

When you think about "Like Attracts," you will be able to see exactly what it means. Your way of life at heart, your outlook and the way you feel, as well as your hopes and goals, are likely to draw similar objects to. The kind of energy you give to the universe is the same type of energy that draws you.

Consider the instances that you were angry at, frustrated, and late. The more angry and frustrated you felt about the day, the further you were running. If you keep focusing on being late, frustrated and angered and frustrated, the more you realize that you give yourself more reasons to be angry, frustrated and late. Consider a pleasant day in your life. A day where everything is going your way. You could be thrilled and content and there appears to be nothing to cause you to feel unhappy. When you focus on the happy and exuberant feelings the more you observe how much you'll be thrilled and happy.

This is the principle behind the Law of Attraction. Like attracts Like. When you focus on positive and positive things and positive things, the more the world provides you with good things as well as positive experiences.

This idea has been in use for a long time however, it's only recently been popularized as growing numbers of people start to realize it is the Law of Attraction is actually Quantum Mechanics, a theory of how the universe works. Quantum Physics is a theory that teaches that there is no set limit and that there are no boundaries as everything is vibrating energy. The energy is in your control and is controlled by thoughts. It's shapeable, formable and mouldable. It's different from simply wishful thinking and hoping. It boils down to faith. To create your Law of Attraction work for you, you have to believe you can trust that the Universe will provide you with the things you truly want.

It is possible that the Law of Attraction could end to be one of the easiest laws you've ever encountered. When you understand it fully and can make use of it, you'll discover that you're able to attain everything you've ever imagined.

It is believed that the Law of Attraction is something that tells people to attract things to themselves by focusing on specific things. It is linked to Quantum Mechanics, which explains that there is no absolute rule and that there are no boundaries. In the view of Quantum Physics, all is composed of vibrating energy. It is believed that the Law of Attraction and Quantum Physics are therefore closely related and are, in fact, they are both interconnected.

Based on Quantum Physics and the Law of Attraction People can be the creators and architects of their world. The universe is comprised of building blocks. It is not as rigid as Newton's classical physics but fluid and constantly changing, similar to Quantum Physics.

It is said that the Quantum Law of Attraction, means that since everything is constantly evolving and changing--and, in actual it is because the Universe is comprised of these constantly changing energies, everything can attract anyone by being focused. The probability that something can happen to an individual is high, if they are focusing on something.

Based on Quantum Physics, every person is a part in the world. This person is focused on certain things and draws attention to them. And, depending on the issues they are focusing on, the objects that are provided to each individual. So, the entire world is influenced by our emotions. It's actually not a thing that's fixed in stone, but something that's flexible and affected by the thoughts of people and beliefs.

For every person, this implies that their desires could turn into reality. All they have to do is concentrate on what they'd like to achieve and what they've always wished for, and they'll be in a position to attract opportunities for themselves more than they they could. In the real world, bringing things to individuals can only be done by allowing them to follow the laws of both Quantum Physics and the Law of Attraction simultaneously while. Concentrating on what you would like to achieve and keeping them in the top of your list is the most effective way to ensure that you're driven to achieve those goals. You'll discover that you'll accomplish the things that you trust the most effortlessly. It's difficult to believe you can get what you want. But this is the basis of the Law of Attraction.

According to the Law of Attraction, we draw attention to everything we think about. When we consider the connection between Law of Attraction and quantum Physics, quantum physics will explain that nothing in this universe is fixed, and there are no limits. Quantum Physics is also adamant that everything matter in our universe contains energy that vibrates.

If you would like to achieve your goals and free yourself from feeling trapped You must be convinced that everything in the universe is energy and that it is in a state of potential. You must accept the law of attraction to take effect for you to be successful. Remember that we are the creators of the universe. According Newton's classical physics the universe is composed of discrete components. They are solid and cannot be modified.

Quantum Physics provides a reason that there aren't separate elements that make up the world. All is composed of fluids and can change over time. Physics visualizes the universe as a deep sea of energy that is constantly appearing and then disappearing out of the universe.

The people of this present world are transforming the energy of the world through their thinking. This is why it is easy to create what one wants to accomplish. In the end, humans are the primary ones responsible for achieving their goals, as well as the destruction of their goals.

The most important thing to grasp is that quantum science has made us makers of all the world. The universe is all energy.

It is imperative that you have heard of the famous Einstein formula. The formula was first discovered in 1905 and reads like this:

$E = Mc2$.

The above formula clarifies the relationship between matter and energy. Matter and energy are able to be changed rapidly. The bottom line is that everything is there is energy and it is constantly changing. Our thoughts have a significant influence on this energy. Energy is easily generated as well as shaped and shaped through our thoughts. It is easy to transform the energy we think into the things we truly desire to be.

Quantum Physics is also referred to as the Physics of Possibility. This theory goes against the conventional belief that the world outside is real, and that the inner world of fantasy. It states that what is happening inside determines what happens on the outside world. The world we live is made by our ideas.

It is impossible to make any changes in the world, as we've mentioned before. It is important to understand that when we focus on our thoughts and the things we would like to see in our minds, we are able to achieve what we desire. But, we must not forget that "it could be done" and it will occur.

It is believed that the Law of Attraction and its powerful relationship to quantum mechanics can allow you to achieve the fulfillment and success of your goals. Be aware that good things can happen to people simply because they believe they're likely to.

The Law of Attraction and Quantum Physics are inextricably linked. It is said that the Law of Attraction notes that by our actions and thoughts our actions, we can manifest our reality. Not surprisingly, quantum mechanics will help explain this Law of Attraction.

The most under-appreciated and unpopular field of science in the present is quantum Physics. Quantum Physics is a deep study of the nature of our existence and attempts to comprehend how micro-influences macro and understand the basis in quantum physics. Law of Attraction.

Although quantum physics is not completely understood due to the deficiency of resources that allow us to look deep enough to understand the entire universe, what we have discovered so far is sufficient to grasp how to apply the Law of Attraction in the realm of thought.

The most important findings in quantum science is the fact that matter can behave like an object or wave. Let me explain the meaning of that. Particles are solid substance. It is able to only be in one location at a given moment which means that you are able to locate its location. However, a wave not an infinite point.

Quantum mechanics has realized, through observation, is that when tiny particles are fired, also known as electrons, through two slits, they behave like particles. Every electron picked the slit, passed through it and then struck the screen's back.

In the end, firing thousands or hundreds in these absorption patterns was the formation of a pattern of two slits. If the electrons weren't recognized when passing across the two slits a large pattern of interference was seen on the back of the screen that is the result created through the wave. The pattern also revealed a pattern of interference caused by the slits. This confirms that electrons went through the slits in waves, and not as solid particles.

What does that mean to us? Our perception emotions, feelings, and perception influences the world around us. When scientists attempted to follow the electron's path to determine where it would travel, they discovered that no matter where the person who was watching desired it to go is the place it would end up. The implications of this are equally significant Our hopes, ideas and opinions literally form the subatomic world all around us!

Evidently, how powerful our beliefs, feelings desire, values, and desires to influence change and create reality is that which we are told by the Law of Attraction tells us. With a knowledge of science, you may be able to shake off the current belief system and test the concept. If, by chance, you were told you could get all you want, wouldn't it be worth giving it a shot? Be amazed and be amazed.

Chapter 4: Quantum Physics And You

We have just entered the age of Aquarius. This signifies we are in the middle of a transition. Solar System has moved into an entirely different location within the Galaxy. The humans living on Earth have never been in this space before. What is to be expected from the new world we inhabit is yet to be determined. When we begin a big cycle there are always hints or clues as well as broad stroke rules of what to expect. We've been in this space for 50 years. There's still another 50 years to take. What this brand new "Day" has taught us is already incredibly important in the direction we should follow.

The first thing to note is that what we call the "Pisces Age" that we've left behind is now over. While some concepts, models, and types that were part of that Age of Pisces continue to be alive and strong the plug on their genesis was pulled. They're no longer connected by the power and power of Galactic Sun. This is because the Galactic Sun has now been driving this Aquarius Age.

The Age of the Pisces was also known as"I Believe" or "I believe." It fulfilled its purpose as humanity was required to construct its beliefs from "Authorities.'

The Age of Aquarius is the age of "I Know." We'll be able acquire all the knowledge that we require to evolve and grow right from within. It's not necessary to have a middleman.

This is possible because of one of the very first gifts offered to humanity by "Day of Aquarius." It's the "Age of Mind,"" the blessing from Quantum Physics is for the mind. These laws will provide guidelines to follow in the coming 1950 years.

They just say that there's the infinite sea of thoughts that is intelligent energy referred to as"the Quantum Ocean. It's actually God's mind. God.

All that's ever existed and is currently or will continue to be in existence there. There is no either present or past. There is only the present moment. There's nothing for you here. There's no width, length No Depth, just the here.

In the real world what we call the Quantum World, the Mind of God is an infinite space called the"HERE-NOW. The way we appear to be a part of the infinite space and also outside of this infinite space in Planet Earth is still a puzzle to our limited minds. This will become apparent to us as we move through the 2000-year time frame. In the meantime, we must to apply to apply the Laws of Quantum Physics and the idea of Quantum Ocean, also known as the Quantum Ocean and that is the Mind of God to create a new reality for us and the world all over the world.

One theory states it is believed that Quantum Ocean, the Mind of Nature, is open to our "Thoughts." Thoughts are the things! It is essential to be more than thinking over things like the Quantum Ocean, the Mind of God, we need to understand how to access it. What can we do to get there. We must learn to coexist 24 hours a day and psychologically within the Quantum Ocean and physically on the Earth. Earth. We must learn how to utilize to apply the Laws of Quantum Physics in the Quantum Ocean to restore our existence and our planet.

Quantum Physics and Your Health

The Laws of Quantum Physics posited that all energy is energy. There is only one fundamental force and it is found inside the Quantum Ocean. Scientists are putting around theories and theories about the primary element of Creation. The term they're using is "Quanta." They're talking about waves and particles regarding possibilities and odds. Words, Words, Words! Definitions are more abstract than what the word itself is.

The word "apple" on the paper isn't an actual tree. It is not possible to consume words like "steak" that is written down on paper. The energy that flows through the Quantum Ocean and "blinks" in and out' of your physical world is the primary element of Creation.

Where did this idea come from? God or, more specifically The Quantum Ocean, is the Mind of God. It's the only thing physical scientists take out of their humor in their rants and blabbering on about Quantum Physics.

They won't be able to come up with an accurate answer until they include God, the Creator God as a factor into their calculation. When it comes to what we're all about and what brought us here and where we're going The 'Big Bang Theory' idea is not the right answers to those questions. It isn't able to answer, "Who put the energy in the ball that created the acceleration in the first place?" God, the creator, God did.

As we learn much more of the Laws of Quantum Physics and the extent and size of the Quantum Universe, the true nature of God the Godhead, God, will be far too vast for us to grasp with our limited minds. We must explore the Quantum Ocean and that is the Mind of God and to work in tandem using the power of the Creator.

The entire Laws of Quantum Physics point at the intelligence of Creation. Quantum Ocean Quantum Ocean is known to be an infinity Ocean of Thought, Intelligence Energy that reacts in response to thoughts as well as our emotions. It's the brain that drives the creation of.

There is a Creator there exists a concept of the Creator. This idea is located in the "mind" of the Creator. Because it is the Mind of the Creator is the Quantum Ocean, there is a plan within that Quantum Ocean.

All that has ever happened, is and is or ever will be, is a part of God's mind. creator, God, the Quantum Ocean. It's manifested in the form of Divine Blueprints, the Divine Archetypes. These patterns are constructed in a manner that when they "blink in the Loom of our physical reality, they take on specific form.

It is believed that there are Divine Blueprints to create the perfect Animals and Men, Women as well as Planets, Birds, Galaxies, and the Universe. If they were first created, these Divine Plans initially 'blinked out in the Mind of God they could be imagined to be an event known as the

Big Bang. It was a fast event. Every single one of us is a unique soul inside God's mind. God. We're "blinking out" and 'in', Birth death, as the universes. There's an Divine timing plan for everything. That's what Bible's reference of "There's an Season" is referring to for everything!

To achieve greater health, we need to balance our current energy system (mind as well as the body) to the initial energy structure of our Soul. It is important to learn how to enter the Quantum Ocean and search for and align with our personal Divine Blueprint. It is time to eliminate all the bad "Dross" that we've picked up in our travels. It's the Quantum Ocean is the place where the blueprints for optimal health exist. It is time to go into and get back the Quantum Ocean and that is, the Mind of God.

Quantum Physics and its connection to Self-Esteem

No matter what your opinions is, it's true that the Universe is full of mysteries that are unsolvable and acts in a manner that is not random. The universe was created by Albert Einstein who said, "All is not randomly," and science is becoming more inclined toward the notion that the universe is acting in a way that is intelligent and there is some source of energy which shapes the universe.

The latest research suggests that our universe made from Dark Matter, Dark Energy (they do not know the definition of it) and Sub atomic Particles that are composed of matter as well as material that fills in the gaps between matter and space, i.e. Space. Space is our Energy!

Sub-atomic particles do not constitute particles by definition They are instead particles of energy that when observed transform into solid objects. This are the stuff we see in the universe and our daily life. Albert Einstein's formulae e = mc2 teaches us that the equation e (energy) =

(energy) x m (mass) (mass) x C (light velocity) is divided by.

This means we that All Mass is equal to Energy. All mass is equivalent to energy and not only sun, gas, oil, or Uranium. All MASS equals Energy. It demonstrates that all of the universe's galaxies, planets, stars moons, gasses, and even the US are energy. We're not made of "we are made of" the energy (there's enough energy within the human body for it to fuel the small village for two weeks).

These subatomic particles do not simply sit in a mass and form matter. they vibrate, float and change shape and form inside as well out, creating and performing the form of a dance directly connected to changes in our observation. They have a mind and are able to move through time and space without a hindrance. They are present both in the present, and the future and in the present moment, they can react and develop on the basis of the information they have that is gathered from all of these dimensions.

They don't make judgments about the right or wrong way They'll formulate whatever you want However, they'll also give you instructions and warnings to help you on your way in the event that you require it. But, obviously it's entirely up to you to decide if you're willing to accept the advice and alter your attitude to think about the messages you're getting. This is known as Free Will!

It is believed that everything that exists already exists, and it is only through our awareness of it (through thinking) that makes it a part of our perception.

Think of Thomas Edison, who performed thousands of experiments in order to discover the light globe to provide an illustration. By utilizing The Law of Attraction and by focussing on the outcome the Edison wanted, Edison knew all he required to do was to persevere and eventually come up with the correct formula to create the globe of light. The light globe was always there as the possibility, but it was only realized after Edison realized what it was supposed to be after hundreds of tests.

Based on the laws of attraction (The Secret) the man knew the time would come when he could reach his goals. How it took so many tests to achieve this could have a lot to be due to the thoughts that were going through his head during that time. It could be because of his ego, doubts, or even his own belief system however, one thing is for certain: with perseverance, he finally attained his goal, and the globe of light was real when the world was finally seen as it was. It was when he finally began to vibrate with the same frequency that he had set for himself.

If we focus on our thinking, feeling or believing in the things we want from the subatomic particles that comprise every aspect of our world even space itself, then we start to collaborate and create it into our lives.

The things we desire are always an option and the job of the universe is to offer it to us and then go around to give us what we desire. It is determined to provide you with the things you want, and as it is, and it's the reason why it exists. Forces that are beyond our comprehension are at play and the sequence of events starts to move to a point that can lead to situations which will provide us with everything we desire.

At some time in our lives every one of us will face challenges. It's unfortunate that the majority of people aren't sure what to do about them and hold on to these issues, regardless of the reason for the issue may be long gone. It's just an issue of creating a situation where it will be a reality in our own experience. It's the same to get what we want or to become who we wish to be. It's already an option however, it's not present until we notice or experience with the same frequency or otherwise you must believe it's that way. If we look into it, believing that it must be true, and paying attention to it and attention, the universe begins to transform into the world we experience.

We're beginning the process by focusing our thoughts, our focus and, as the time-frame needed to construct what we envision is in place as it changes to reflect our current reality. In the same way, for us to create our own reality, we must consider it first. The thought process is like an outline that the universe makes use of to create our world. So, if we wish to make our lives better or attain what we want we must think or communicate our thoughts to the universe in order that it will construct our new reality.

It's time to create a brand new line of blue prints to provide us with what we would like!

All is possible to exist so, which means that the possibilities for your future, in the way you envision it can actually be realized. It's a matter of living your life the way you wish to live it or operating at the same speed in the way you wish to, and the world will, with time, manifest into the way you perceive it and eventually be your reality.

Self-esteem issues and distrust are both negative emotions, and are the means to hinder the positive manifestation. Anyone suffering from these issues concentrates on the negative aspects of life rather than what is good or right in their lives. They don't understand the reason why their lives cause such a lot of suffering. They think that by feeling sorry for themselves, the entire world should feel bad for them. They are able to spend their entire time with self-pity and hope they will be able to alter their situation and provide an amazing solution to their plight however, they do not respond.

The less they react and the less they respond, the more they feel regretful for themselves, and the downward spiral goes on to spiral downwards. Only you can make a difference in your life and the unhappy feelings you feel. We have demonstrated that the world will give you what you want , and offers it to you in abundance. It builds your world within your headand will continue to do this until you tell it different.

If you are suffering from a anxiety or lack of confidence, you are not alone. self-esteem, your thinking is what causes your struggles throughout your day. We've found that the universe creates and builds what you desire with the thoughts you think of as blueprints. The more precise and clear your thoughts are, the better and faster you'll be able to create your dream. Drawing the blueprints with the positive emotions as a guideline and the universe begins to construct from the moment the pen is placed onto paper, so say.

It predicts what you'd like using your previous experiences, as well as future ones and starts to build them all according to your messages in the present. We all live in the present and the past... In other words, I mean, we're living out the results of the thoughts we think, words, and our actions from our past (in another way, we are living the numerous past yesterdays). If your thoughts and words were joyful, happy positive, clear and unambiguous yesterday, this is the kind of life you're likely to live today. If your actions and thoughts yesterday were filled with regret, anger, sadness, regret anger, discontent or insufficiency, it's the way you'll have live today.

It's impossible to change the past or your present, but you can alter the future you wish to, today!

The universe is not able to distinguish between positive or negative thoughts. If you concentrate your mind to what you desire with certainty, confidence, and faith, you'll be granted it. If you focus your attention on the things you don't desire with conviction and honesty then you'll receive the opposite. It's vital to know that. Your thoughts, no matter how positive or not, are utilized as directions (blueprints) to create or assemble the things you would like, no matter if you'd like it or not.

Self-esteem issues and distrust are both negative emotions, and are the means of hindering positive manifestation. Anyone who is suffering from these issues concentrates on the negative aspects of life, instead of focusing on what is good or right in their lives. They are unable to comprehend the reason why their lives cause such suffering, and they think that because they feel bad for themselves, the entire world should feel bad for them.

They remain with self-pity, hoping that the universe will alter their lives and offer some magical solution to their discontent. They believe that "This is the way I am living my life, and I could prefer to settle with less." They tell them, "I'm not good enough," and in doing this, nothing alters and the cycle goes on.

Only you can alter your life and the circumstances that cause you to be unhappy starts from within. As we've seen in the universe, the use of subatomic particles, it gathers the things you desire, regardless of regardless of whether you are conscious or not. It offers all of it. It creates an image of the universe within your headand will continue to do this until you tell it to do something different.

If you're suffering from a lack in trust or self-confidence, ask yourself "What kinds of thoughts could make this become a reality?" Be conscious of your thoughts and utilize your emotions as indicators. If you are feeling down this tells you not to take that route. If you're feeling great it says this is the direction you should choose. Be confident! The universe has provided you with the tools you need to navigate.

Don't worry about it even if you don't succeed or the process doesn't go at the speed you desire The universe will give the perfect solution to make progress even if it seems like a backwards step. The nature of life is more and it's in opposition to nature's forces to diminish the quality of life. If you believe that you're going backwards, but you're actually moving forward. all you have to learn is from your mistakes.

Take a look at your life as if an audience member rather than the main character in the movie. Take a look at the mistakes and try to figure out solutions to correct them. The word implies that they're only missing, like scenes in a film They might be annoying in the moment but all we have to do is return to the scene, and then move onto the next. Life is like a film in your head, so ensure that it is a positive one.

Chapter 5: Aspects Of The Building Blocks Of Matter And Wave-Particle Duality

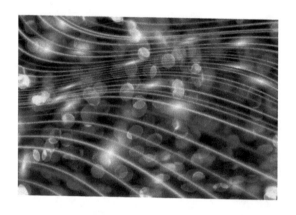

One of the most frequently asked questions is 'What is the basis of things what? The majority of the answers been given on the basis of having a basic set of elements that make up everything. The idea was first studied by philosophers, later by alchemists, then the chemists. It was a physicist's job to discover the answer!

I'm certain that you've heard of one of the very first models that we have heard of. The entire structure was an amalgamation of four elements: earth air, air, fire and water. It was advantageous being able to build only a few blocks and connections between the contents and properties.

The second major model was Mendeleev's chemistry periodic table. Everything was made up of atoms. There was one kind of atom for each element (e.g. iron, oxygen). More than 100 components have been identified, a vast majority of them are simple elements!

Take a gold coin along with you. Continue to cut it to smaller chunks. Every lump will always turn out turn out to be pure gold. In the end, it was believed that you could reach the point where you can't cut it anymore. The early Greek word "atomos" (meaning "uncut") is used to describe these small pieces in an element. The word is a reference to indivisible, meaning that the tiny lumps were believed to be essential.

But, physicists have found that electrons are derived inside the atoms. This meant that the particles had to be composed out of something other than water. The hunt was on. Rutherford looked into and studied the Plum Pudding System of JJ Thompson at the beginning of the 20th century. It was believed that electrons with negative charge were kept in a positive "batter.' He asked two pupils, Geiger as well as Marsden for a test to prove the theory.

Certain things we're glad to refer to as things; chairs, tables and squirrels. Certain things that we're content to call sound waves ripples in water, Mexicans. Certain things, however, caused long debates over what they're. Light is an important instance. At the close and beginning of 17th century Newton believed that light was composed small particles (he identified them as corpuscles) and, consequently, matter. Huygens believed that light waves were. At the time, Huygens won the debate through a series experiments that showed how light behaved as a wave and how it could travel across the gap (diffract). The problem was solved before we began to explore the subatomic world.

It was 1905 when Einstein released a study on a dark phenomenon dubbed "photoelectric effect.' It was found that when light reflected off certain metals with electricity, they lost the charge. This wasn't the case for all metals however, nor the entirety of light. Zinc is a good example. It retains its charge when white light hits it but is depleted as ultraviolet light (the type employed to create tanning booths) hits it. It is not clear in the event that light is an electromagnetic wave. Einstein recognized the evidence showed that Newton was right and that light actually was made up of particles. He called them the 'quantum of light', or "photons.' An area of Physics known as quantum mechanics has emerged.

But wait a second, Huygens' experiments have demonstrated that light is in fact a wave. The experiments are still in operation even today. What's the matter could they be both correct? Actually, they are. It is apparent the light wave is both as well as particle. It's behavior is based on the conditions. Do you recognize it? Yes, it's similar to mass-energy as well as space-time. In this instance the concept is called "wave-particle duality," and we believe that light is composed of wavicles (from WAV-PARTICLES). Therefore, if we believe that light is an element and it turns out to actually be waves what would we do with the things we were convinced were particles?

The year 1906 was the time that JJ Thompson received The Nobel Prize for proving that electrons are particles. He was able to prove that they had quantified their mass and charge, which was in fixed lumps , instead of being able to possess any quantity. His co-worker, George Thompson, was awarded the Nobel Prize for proving that electrons are waves. We now know that both are wavicles and both Thompsons proved right.

Quantum Physics - The Discovery that Scientifically Demolished Materialism

Quantum physics has shown that subatomic particles are able to appear and disappear in the vacuum. The idea that matter could be a result of quantum level, it is a property that is related to matter. Some scientists attempt to understand the source of matter and its non-existence at the beginning of time. the property of matter and show it as a an integral part of the laws of nature. In this model, the universe is represented as a bigger subatomic particle.

But, this syllogism absolutely not feasible and will not in any way describe why the universe began to form. William Lane

Craig, author of The Big Bang: Theism and Theism is adamant about the following reasons:

Quantum mechanical vacuum that spawns particles is not the conventional notion about "vacuum" (meaning the absence of). Quantum vacuum is a sea continuously forming and dissolving particle which draw energy from vacuum to sustain their short existence. It's not "nothing," and therefore the particles of matter don't arise from nothing.

In quantum mechanics, the matter can't function as it did previously. The result is energy transforms into matter and, then, all of a sudden the energy vanishes. In short, as it was stated there is no condition of existence from nothingness.

As per Isaac Newton, light was the movement of a substance called corpuscle. The basis of traditional Newtonian physics-which was widely believed prior to the time quantum physics was discovered, was that light was particles. But, James Clerk Maxwell, one of the 19th century's physicists who believed that light could be a form of waves. Quantum theory has solved the most controversial issue in the field of physics.

The year 1905 was the first time Albert Einstein claimed that light contained quantum, or tiny particles of energy. These energy packets are known as photons. While they are described as particles, it can be seen that photons behaved according to the wave motion suggested in the work of Maxwell in the early 1860s. Light therefore was an oscillation between particle and wave which was a situation that revealed a major contradiction in Newtonian physical physics.

Following Einstein, Max Planck, famous German physicist, looked at light. He stunned the entire world of science by concluding that it was an electromagnetic wave and a particle. In the wake of this theory that he developed under the name quantum theory, the energy is dispersed as a series of interspersed and discrete packets, instead of being straight and continuous.

In a quantum phenomenon, light exhibited particle-like and wave-like properties. The particle, also known as the photon has been observed as a wave in space. Also light moved around space in the same way as waves, however, it behaved as the active particles when it came in contact by an obstacle. To put it in another way it was a form as energy, until it encountered an obstacle. At that point, it changed into particles, as if they was composed of tiny bodies of material that resembled of sand grains.

This theory was further refined following Planck by scientists like Albert Einstein, Niels Bohr, Louis de Broglie, Erwin Schrodinger, Werner Heisenberg, Paul Adrian Maurice Dirac and Wolfgang Pauli. All of them were recognized with Nobel Prizes. Nobel Prize for their discoveries.

In relation to this discovery regarding what is the essence of light Amit Goswami writes:

When light is perceived as a wave, it appears to be able of being located in 2 (or more) locations simultaneously, as when it travels through the slits of an umbrella, it creates an diffraction pattern. When we photograph it on photographic film, it appears discretely in a spot-by-spot manner as particles. Thus, light has to appear as both waves as well as particles. Paradoxical, isn't that? One of the mainstays of physics from the past is at stake: a precise description of the language. The notion of objectivity is also in question How does the character of light, or the elements that light depends on, affect our perception of it?

Scientists are no longer assuming that matter is made up of random, inanimate particles. Quantum physics did not have any material significance since the fundamental nature that matter had was mostly non-material. When Einstein, Arthur Holly and Philipp Lenard Compton were investigating the structure of light's particles, Louis de Broglie began to study the structure of its waves.

De Broglie's discovery was truly remarkable and he discovered during his research that subatomic particles typically had wave-like properties. The proton and electron were able to produce wavelengths. That is, inside the atom, that materialism portrayed as a total matter was a non-material waves of energy contrary to what materialists believe. Like light, these tiny particles within the atom could at times behave as waves and displayed the characteristics of particles found in other. Contrary to conventional standards absolute matter within the atom was visible at specific times, and disappeared at other times.

This significant discovery has demonstrated that the way we think of is the real world, actually shadows. Matter had finally left the realm of physics, and was moving towards metaphysics.

Now we understand how electrons and light behave. What is it? If we claim that they behave as particles, then we give an incorrect impression, even if we claim they behave as waves. They behave in their own unique way, which can theoretically be termed quantum mechanical. They're doing things in a manner unlike anything you've ever seen before. Atoms don't behave as a weight which hangs from an elastic spring and moves. Also, it isn't an image that of solar systems, with small planets that orbit. It doesn't appear to be fog or mist or something similar, which is in fact that is circling around the nucleus. It's unlike anything you've witnessed before.

There is at minimum one simplicitation. In this sense electrons behave exactly the identical way that photons behave; they're both dirty but they behave exactly in the same manner. Therefore, how they behave requires an enormous amount of imagination to understand, since we're going describe something different from the ones you've ever heard of. We don't know what that might be.

In short quantum physicists claim that the world of reality is a fantasy. The most famous quantum physicists from the 1920s all the way starting from Paul Dirac to Niles Bohr as well as from Albert Einstein to Werner Heisenberg tried to explain these outcomes of quantum experiments. Then, a small group of physicists came to an agreement referred to as the Copenhagen Interpretation of Quantum Mechanics during the Fifth Solvay Conference on Physics that was held in Brussels in 1927. Bohr, Max Born, Paul Dirac, Werner Heisenberg and Wolfgang Pauli. The name is derived from the home of the group's who was its leader, Bohr, who suggested that the physical reality suggested in quantum theories was the data we possess regarding the system, as well as the conclusions we draw from this information. In his view, these notions we make in our minds did not have anything to do with the reality outside.

In the end, our inner world was not in any way connected with the external world which had been the focus of physicists' research since Aristotle until the present. The physicists discarded their previous ideas regarding this notion and realized that quantum understanding only represented our understanding of physical phenomena. The world that we perceive is only information stored within our brains. Also we will never experience direct experience of the material world outside the world.

One of the most crucial and most dangerous assumption we learn during our childhood is the notion is the existence of objects that exist beyond -- and independent of the subjects who are their observers. There is evidence to support the validity of this notion. When we gaze at the sky, as an example we think about the sky in the same way we imagine to see it, in the classically defined course. It is natural to imagine that the moon remains present in space-time when we aren't looking at it. Quantum Physics is saying no. If we're not looking to it, moon's probabilities of expanding wave is increased, but only by a tiny amount. When we gaze at it, the waves disappear instantly, since the wave cannot be within space-time. It is better to apply the metaphysical idealistic hypothesis that there is no object in space-time that is not an individual who is aware of it. This holds true, naturally to our world of perception. The existence of the Moon is evident to the world outside however. But when we think about it all we can really feel is our own perception about the Moon.

The role that observation plays in the field of quantum physics can't be undervalued. When it comes to Classical Physics

[Newtonian Physics] the structures that are observed have an uni-personal nature that observes and investigates the structures. Quantum mechanics however, only through the process of observation can a physical thing be considered to have a real worth. What causes things to happen isn't more objects. However, it's concepts, ideas and information that create things.

After the most delicate and intriguing experiments that the human mind could have created over the course of eighty years, there are no opinions against quantum physics that has been scientifically and conclusively proved. There is no challenge that can be put forward to the conclusions made from the tests conducted. Quantum theories have been studied in many different ways by scientists. It has been awarded the Nobel Prize for a variety of scientists and continues continue to do it.

Matter, the most fundamental idea of Newtonian Physics, which was once believed in absolute certainty, is now gone. Materialists, who were the advocates of previous conviction that matter is the sole and only element in existence, were actually puzzled by the absence of substance that was suggested by quantum Physics. They are now required to define every physics law in the field of metaphysics.

The shock that materialists suffered in the early 20th century was much greater than the one discussed in the pages. However, quantum physicists Bryce DeWitt and Neill Graham have described it this way:

There is no breakthrough in modern science that has had more impact on the human mind than the rise the quantum theories. Dissolved from the ages-old paradigms of thinking, the scientists of the past were forced to adopt the new metaphysics. The stress that resulted from this change of perspective persists until today. In essence, physicists have suffered the loss of a substantial amount of their control over reality.

Chapter 6: Quantum Possibilities And Waves

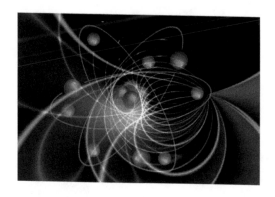

The quantum theories explain how probabilities are determined by looking at all possibilities that could be involved in any given process. If there are multiple possibilities to consider Quantum theory can tell us how to mix the various possibilities in order to discover the possibilities that result. The probabilities of a certain outcome of a measurement can be calculated from the possible outcomes. Potential quantum effects for matter like electrons behave in a certain way similar to waves.

However, the electron's only particle therefore, what's the meaning of 'waving'? In general, we view the term "wave" as being a motion in a physical medium that extends such as sound waves in the air or ripples that appear on the lake's surface. This same definition for physical wave can't be applied to a single electron that, when discovered, is determined to be located at a certain level that is not scattered within a specific region. However, the concept of a wave is applicable to single electrons since it accurately explains how quantum possibilities that are associated with various measurement results shift with time and shift across the space.

What are the wave patterns?

Waves are a series of patterns orchestrated across the space. When a rock is dropped into a lake and it causes ripples across the top of the water. These ripples move away from the rock-hit place. The ripple pattern moves across the lake, and the water molecules oscillate in its own predetermined positions. The molecules fluctuate between rising and falling and cause the surrounding molecules to also rise and lower in a slightly different time frame from the movement of the surrounding molecules. This synchronized motion in water molecules is a contributing factor to an exchange of momentum and energy across the surface. The duck floating a certain away from the source of the wave (the rocks entry point) is influenced by this momentum and energy and can oscillate between up and down. Be aware that the flow of energy and momentum is not a part of the actual flow of water between the source of the wave and the place where the effects of it will be perceived (the water duck). A sample of a wave is illustrated in FIGURE 1.0 as the pattern of a moving. Locations with the highest wave height are referred to as 'crests' while low wave height locations are called 'troughs.' The pattern formed by two crests that are adjacent (or two troughs) is known as a complete cycle. The distance between two crests adjacent to each other is"length" or the duration of the complete cycle (also known as the "wavelength"). The wave or pattern tends to move in a direction that is forward-looking, however there aren't any other objects that move in the direction in which the waves move.

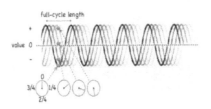

Image 1.0 An eddy is illustrated to be of a certain magnitude (wave height for the water wave) which oscillates both positively and negatively in a normal way. The space between the crests is what is

the duration of whole period. The amount of time that passes during a entire cycle of oscillation equals the entire cycle.

A point (or the duck) of the wave oscillates upwards and down in a typical frequency known as the full-cycle (also known as the "phase" oscillation). The figure illustrates the clock that keeps time as a particular spot on the wave oscillates between up and down. The speed at which the hand revolves around the clock face is determined by the amount of spring of the wave medium like air or water and the density. If an internal timer of the device is set to zero, the pattern appears as illustrated by the darkest curving of the line shown in FIGURE 1.0. As the clock's hand turns it shifts the pattern in a smooth direction to the left, as evident by the series of lighter curly lines. Each time the clock rotates around once, the wave moves at a distance that is that is equal to the length of the entire cycle.

The speed of the wave is the same as the amount of time divided by. This is known as the speed of the wave (wave velocity = full-cycle duration/full-cycle duration. The relationship between full-cycle length and full-cycle time as well as wave speed can be illustrated using a cartoon-like style.

a fictional ladder, as in the figure 1.1.

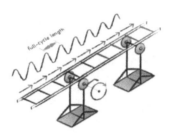

Figure 1.1. The mechanical explanation of how the full-cycle relationship is related to duration full-cycle length, full-cycle time and wave speed.

Between the ladder's rungs represents the complete distance of the wave. The ladder is anchored by the mechanism that is specifically designed to move it by using an arrangement of a crank and wheel. The wheels that are attached to the crankshaft grip the top part of the ladder via friction. As the crank rotates and the ladder is moved in the direction indicated by the arrows. The wheel's diameter is such that , if you rotate the crank at a single time the ladder moves at a speed that is equal that of how far the rungs on the ladder. If you rotate your crank in a consistent pace, the ladder will be constantly moving forward at a constant rate. For instance, if rotate the crank every second, and each rung is separated by 1 foot and it is said that the rate of climb will be one foot every second.

What exactly is Wave Interference?

When two rocks are thrown in a pond, each creates waves of ripples which move away from the point the rock is struck. If a duck is floating on the surface that is shallow then the force of both waves may be felt. In some areas of the water, effects of both waves reinforce each other, leading to an oscillation that is broad and up-and-down (the duck is on a wild ride). The effect of this reinforcement is known as positive interference. At certain points within the lake, the behaviour of both waves could occur in reverse, and cause no oscillation between up and down. (The duck remains there.) The canceling effects are known as destructive interference.

Chapter 7: Application: Quantum Computing

Does information have physical properties?

Computers are machines that process information. Computer scientist and physicist Rolf Landauer, argued that knowledge is an integral component of our physical universe. He explained this as follows the idea that information isn't an abstract, disembodied entity and is always tied with physical manifestation. It is represented through engraving on a stone tablet turning it magnetically or charge or a hole in the card that has been punched or a mark on paper or other alternatives. This connects the handling of information to all possible possibilities as well as limitations in our current physical world as well as the laws of Physics as well as its storage of parts. In the event that "information is physical" as Landauer claims, then it is necessary to handle it mechanically. That is the physical method for storing information and processed by computers must be examined with the help of quantum theory. It aids in understanding the nature of computation before looking at quantum computers.

What is a computer?

The term "computer" refers to a computer that takes and stores input, processes that information in a programmable order of operations, and generates the resultant output of information. The term "computer" was first coined first in the 1600s to those who do computations or calculations. It today, computers perform computation. Computing machines can be classified roughly into four categories:

1. Computing equipment for classical Physics. These machines make use of moving components, such as gears and levers, for computation. They are usually not programmable, but they always do the same thing for example, like adding numbers. One instance could be that of the 1905 Burroughs Adding Machine.

2. Electromechanical classical mechanics that can be fully programmed computers. They function through electronically controlled moving components. They process information stored in digital bits, which are represented by positions of a variety electronic switches.

The first of these machines was constructed around 1941, in 1941 by Konrad Zuse in wartime Germany. The theory behind their programming can solve any issue that could be solved using algebra. They were the first "universal computer in this regard.

3. The hybrid, electronic, and all-electronic classical computer systems for physics. They are fully programmable and universal computers do not have moving mechanical parts , and operate by using electronic circuits. The first computer to be developed was the ENIAC designed by John Mauchly and J. Presper Eckert, University of Pennsylvania 1946. The physical principles that explain the motion of electrons inside these circuits are based on quantum Physics. In the event that there aren't any superposition states or entangled states that involve electrons in various circuit components (capacitors transistors, capacitors, etc.) The classical physics describes the way in which electrons are used to represent information. Therefore, we call these machines--essentially any computer in operation today--'classic computers.'

4. Quantum computers. If they are ever developed successfully they will operate in accordance with the rules of quantum Physics. The quantum state of knowledge will be expressed in their quantum state of electrons, or other quantum artifacts. Additionally, there will be entangled states of electrons within various circuit components. Computers are expected to be able to solve these sorts of issues much more quickly than the conventional computer of today could do.

How do Computers Work?

Computers are able to store, manipulate and manage data by using an alphabetic binary language made up of two symbols: 0 and 1. Every symbol that is 1 or 0 is called a bit, which is short for a binary digit , because it could have one of 2 possible meanings. A text page like the one you're reading is stored in a computer as a number of numbers. Each letter is represented as the binary code. For instance"A" is 01000001, 'B' is 01000001 and so on.

A typical laptop computer every bit is represented as the amount of electrons that are stored in a tiny device known as capacitor. One can imagine the concept of a capacitor as being a box that houses a specific number of electrons. It's like a bulk grain container in a supermarket which holds a certain amount of rice. Every capacitor is referred to as a memory cell. As an example, capacitors could be able to hold 1,000 electrons. When the capacitor's fully or close to being completely filled with electrons we can say it is one. In the event that the capacitor's contents are empty, or nearly empty, that it is a bit of 0. The capacitor cannot be allowed to fill up half-way and the design of the circuitry ensures this doesn't happen. By combining eight capacitors, all of which is either filled either empty or full, an number with eight bits -for example, e.g. 01110011 -- can be read.

The function of machines circuit is to fill or empty diverse capacitors in line with the rules of the program. In the end, the process of filling and emptying capacitors allows it to complete the desired calculation , like, for example, to add two numbers of 8 bits. In a computer, these actions are executed by tiny computer circuitry known as logic gates. Logic gates are made up of silicon, along with other elements which are arranged in a way so that it blocks or transfers electrical charge, according to the electrical environment it is in. Inputs into Logic Gates are bit-values, which are that are represented as a complete capacitor (a 1) or an empty capacitor (a zero). (The term "gate" refers to the fact that something is inserted through it, and then something emerges through it.)

How tiny is one gates logic be?

In the very first electronic computers such as the ENIAC constructed during the late 1940s the only gates was used as a tube made of vacuum similar to the tubes for amplifiers that are being used in the vintage electronic guitar amps. Each tube is at least about the same size as your thumb. In the year 1970 the microcircuit revolution was capable of reducing the size of every gate to less than one-hundredth millimeter. If things become smaller than that is best to determine the length of the unit, which is called the nanometer, which is one millionth of millimeter. The gate's size in the year 1970 was 10,000 nanometers. In contrast the silicon atom, the most important element of atomic chemistry in computer circuitry, measures approximately 0.2 nanometer thick. In 2012, single gates used in conventional computer systems had been reduced to be separated by just 22 nanometers -- which is about 100 atoms. The working size of the gate is just 2.2 nanometers, which is about 10 atoms thick. The small size of the gate permits you to put several billion memory locations and gates within an area that is the equivalent to your tiny thumbnail.

Gate sizes that are smaller than these dimensions can lead into both an ailment as well as the possibility of a blessing. We are leaving the realm of multi-atomic physics, and we enter the world of single-atomic Physics. There are now differences between the classic physics principles that explain the typical behavior of many atoms, and the quantum physics principles which are necessary for single atoms. It's a random-action domain , which isn't exactly appealing in the context of trying to build an appropriately controlled system to perform our mathematical bidding. In reality, a team of researchers headed by Michelle Simmons, director of the Center for Quantum Computation and Communication at the University of New South Wales, Australia, constructed a gate that is made up of a single particle of phosphorus embedded in the silicone crystal tube. It is the smallest gate ever designed. The gate will only function only if it is chilled to an extremely low temperature of 459°F (273 ° Celsius). If the material isn't at the minimum as cold as it should be, this random (thermal) movement of the silicone atoms inside the crystal reduces the encapsulation of the electron psi waves that may escape out of the channel which it is supposed to be contained. For everyday desktop computers that, after all require operating at temperatures at room temperature this leakage can prevent single-atom gates from serving as the base of technology that everybody is able to use. However these tests show the possibility that computer systems can at the very least theoretically, be constructed at the atomic scale which is where quantum physics governs.

Can we design machines that employ the fundamental quantum properties?

Since physics determines the final behavior and effectiveness of data storage, transfer and processing, it's appropriate to inquire about what quantum physics does to play an important role in the field of the field of information technology. Since electronic computers are based on the behaviour of electrons as well as communication systems rely on the behaviour of photons - both elementary particles, it's not surprising that the efficiency of technology for information is affected by quantum Physics. However, there is a nuance. Current computer technology do not use quantum superposition states in order to convey information. They employ states that could be described as classical physical states such as electron groups.

The main issue is how do we build computers that utilize quantum mechanical systems to increase our capabilities to solve problems in the real world? If computers of this kind were developed that way, they could be able to bypass specific types of encryption techniques for data more quickly than any computer currently running currently. This would transform the world of privacy and confidentiality on computers as well as the Internet. An encryption code that could require many thousands many years of time to break on traditional computers could be cracked in a matter of minutes using quantum computers.

What is what is a Qubit?

The term "bit" can refer to the abstract, unembodied mathematical notion of information as well as the physical object that contains the information. It is evident that in the classical physics, physical bits carry an abstract piece of knowledge. There is a simple one-to-one relationship between the state in the physical part and that of an abstract piece. It is either 0 or 1. There is also the possibility of using quantum artifacts that are individual, like an electron or photon, in order to create the corresponding portion. In this instance the fundamental physical object is known as a qubit. It is which is short for quantum bit.' Qubits have two distinct quantum states, for instance the polarization H and V that the photon has, or the higher route and the lower for the electron. If measured, the results indicate a bit number of either 0 or 1. But keep in mind that we could choose different polarization measurement methods like the H/V and D/A. The outcomes could then vary, and the likelihood of observing different outcomes based on the measurement method we choose. In this instance there isn't a one-to-one connection between the status of the qubit's physical state and the significance of a abstract concept. Quantum physics concepts provide significant differences between the nature of qubits and classical bits. Classical bits are able to be copied multiple times as many times as you want, with no damage to the information. qubits are not able to be copied or copied even once, though they may be transported. The status of the classic bit 1 or 0 can be determined with a single measurement. The quantum state of a single qubit cannot be established by any combination of measurements.

What physical principles distinguish quantum and classical computers?

There are a lot of differences in the types of gates utilized in classical computers and those that are needed to build quantum computer. Classical gates carry out functions that are not irreversible. Understanding the output doesn't provide information about which inputs to use. However the quantum gate is going to function correctly with qubits, then it should be irreversible. In other words, you need to be able to identify the state of the input by analyzing the output state. This is due to the fact that every quantum gate operation has to be a unit operation. Remember that we make use of the term "unitary" to describe physical phenomena or processes which cannot be separated into distinct steps with specific, identifiable results. The main example we looked at is the electron (or photon) traveling from its source to its final destination in a situation in which two distinct paths could be taken. We noted that if there's no permanent trace left behind by the electron's journey that suggests that it's chosen a distinct path, it's wrong to claim that it has in fact taken one route or the alternative. It is also incorrect to claim that both paths are pursued. The entire process of leaving one spot and getting to another has to be considered as an unison whole process , which is a single operation. The processes can be reversed.

What logic gates could quantum computers make use of?

Since the beginning of the 1990s scientists have been pondering how a universally-programmable computer built around quantum superposition or entanglement might be created and to what types of problems it might be ideal. This isn't a straightforward problem, and it hasn't been resolved completely to date. On the other hand significant progress has been made and it is likely are reasonable to believe that a quantum machine could be a reality in, say 10 or 20 years. Quantum computers take qubits as inputs and performs various gate-related operations based on an algorithm designed by the developer, then outputs the altered qubits. The output qubits should be at least the same as the number of input qubits in order for the entire procedure to be considered unitary. Like with traditional computers, there exist various options that the designer can choose for the gates that will be utilized in quantum computers. This article will focus on a set identical to the logic of XOR and as described earlier for conventional computers. Quantum computers use the logic I refer to as"QXOR", QR. Utilizing two quantum gates named "quantum XOR as well as quantum ROTATE', the possibility of a universal quantum computer could be constructed, at the very the very least, in the principle. What exactly are these two types of gates?

The first thing to note is that we assign the two qubit states that are possible the names of 0 and 1 and are represented, for instance by the H-polarized and V-polarized states in a single photon. The operating principles that govern quantum computers are not dependent on the way we depict the qubits in physical objects.

Quantum XOR gate or QXOR gate is illustrated in Figure 2.1. With the QXOR gate the B qubit is moved unaffected throughout the gate (from left to right) instead of being removed as in the conventional model. This means that the QXOR gate reversible as well as unitary. In other words, outputs are linked uniquely to inputs.

One method to think of QXOR gate is to consider it as a logical process. QXOR gate could be to think that qubit B is the one who controls the actions of qubit A which is evident by the arrow that points between B and A. In the event that qubit B's status is (0) and qubit A is able to pass through the gate in the same way as shown in sections (i) as well as (ii) from FIGURE 2.1. However, if B is located in the state (1) and qubit A is in state (1), A's state changes. B is altered to (0) (or (0)) to (1) and between (1) into (0) as shown in Figure 2.1 section (iii) as well as (iv).

Figure 2.1 I Quantum XOR gate. Qubit B is the one controlling the modification of the qubit A. Qubit B controls the process of qubit A modification. Qubits shift from left to right.

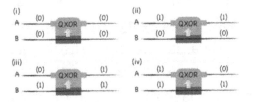

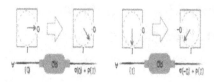

The other type of logic gate needed is the quantum ROOTATE gate, also known as QR gate, which is shown in Figure 2.3. The gate is a single input and an output and is reversible and unitary. When the state of input is (0) and the output is the exact as the superpositions of (0) as well as (1) states with the state arrow pointed towards the direction diagonal. In the case of (1) then the output state will be the same superposition of (0) as well as (1) state, however having the state arrow pointed towards the opposite direction. Both 'a' as well as 'b the arrow components have the value in the range of 0.707 (i.e. the equivalent of one half the

116

square root). Based on Born's Law this means that the probability of getting either, (0) or (1) by measuring each at 0.5 or fifty percent.

Figure 2.3 The quantum-ROTATE gate generates an overposition between (0) as well as (1) state of the qubit. The arrow-rotation diagrams look similar to the diagrams of polarization states.

The two possible output states are like the diagonal and anti-diagonal forms of Polarization. If the qubits are represented using an image Polarization The QR gate can be simple to implement with a crystal that rotates the polarization the arrow by 45 degrees in the clockwise counter-clockwise direction. Quantum computing researchers have demonstrated that they can combine the sequences of

QXOR gates as well as QR gates. all quebit-based calculations can be made.

What is the operation of Quantum Computers?

The classic machine operates by defining input data in the form of bits selection, and then sending the bits into the processor, where gates operate continuously according to the programme and reading bits' values at output.

The quantum computer operates by providing input data in form of qubits, each one with their quantum state defined by sending these qubits through the quantum processor so that gates perform actions on them in accordance with the program and then analyzing the qubits' quantum state at the output. The primary difference between the classical and the quantum scenario is that there can be only overlap and entanglement when in the quantum situation. Quantum states can only occur within the circuit if the operation of all the gates in an integral process. This is only true when there is no possibility to, even in theory that a person could be aware of any one of the qubits (0 and 1) in the inside of the circuit.

What chemistry and physics issues are quantum computers able to solve?

The story of quantum computing didn't start with computer science, rather, it began with Physics. The year 1981 was when Richard Feynman, one of the most creative theoretical physicists discovered that the most fundamental quantum theory equation -Schrodinger's equation cannot be efficiently solved using ordinary computers. Schrodinger's equation is a key element in quantum theory similar to Newton's motion laws found in classical theorizing of physics. The difference is that when Newton's laws define the way that classical objects behave in terms of precise and completely predictable results, Schrodinger's equation explains how quantum states change over time. It is important to note that quantum states do not exist exactly in sync with results of measurements, but instead provide only the possibility for outcomes. The reality that Schrodinger's equation is not able to be solved with conventional computers is a major barrier to the development of science. There is a fundamental equation which we have to solve to predict the probability of outcomes from experiments. However in the event of complex situations that involve a lot of quantum objects, we are unable to solve it! We don't really know what the theory says which is why we cannot utilize it to the fullest extent to improve the field of engineering, science and medical research. We aren't able to develop more efficient quantum-based drugs since it's impossible to solve Schrodinger's equations in the case of large molecules. Naturally, scientists have numerous ways to find the approximate solutions to Schrodinger's equation. This can be extremely useful, however, there aren't any precise solutions that could provide welcome unexpected surprises. Feynman thought that a new kind of computer called quantum computers could be able to solve Schrodinger's equation in a speedy manner. Since Feynman has pointed out this that a great deal of effort has been done to create a computer similar to this. The computer would function using quantum principles instead of the conventional physical principles, which regular computers work. Contrary to the majority of computer science and math issues, the ones with Schrodinger's formula can be transformed into algorithms that are executed by the quantum computer. This is because Schrodinger's equation is the primary equation of quantum theory!

For instance, Schrodinger's equation permits us to determine the energy and the shape of the psi waves for every possible quantum state for the electron inside the atom. Molecules like DNA, the most crucial molecule consist of atoms that are arranged in ways that give them their shape and structure. They also allow them to carry out their duties including encode and transmitting the person's genetics. Since DNA molecules are made up of so many quantum particles, including protons, electrons, and neutrons when in the case of computers that are classical it is difficult to accurately comprehend and determine the nature and purpose that Schrodinger's equation has. To comprehend the reasons why Schrodinger's equation for DNA molecules when that is solved using a classic computer is so challenging to solve, think of an illustration. Let's suppose that the molecule is composed of at least 500 electrons. Actually, it is a small protein molecule when as compared to DNA. To show the quantum states of the five hundred electrons, using computers' bit states, it's essential to show the possible states of entanglement for the fifty electrons.

Each of these states is a quantum possibility for a specific set of outcomes that could be observed if tests were conducted on every electron. To simplify things we'll say that every electron is in two states, identified as either 0 or 1. In other words that it could be considered the qubit. There are four possibilities when the electrons are in two states: 01 1, 01, 10 and 11. There are eight possible variations when you have three electrons. 000, 001, 010 101, 011 110 and 1111. There are approximately 2500 or the possibility of 10150 combinations for 5100 electrons to be thought about. This is a lot more than the totality of elementary particles that exist in the universe! Every combination has to be represented by a specific number within the computing memory. However, it's impossible to store all these numbers on any computer smaller than that of the totality of the universe. The solution might be to break up all the combinations into smaller groups , and treat each group separately by shifting the numbers into or out of memory in the system. However, the time it takes to accomplish this task is likely to be more than the duration that the universe has. This illustration illustrates Feynman's primary argument: as the size of the quantum problem that needs to be solved increases as does the size of the computer required to solve it gets more quickly, exponentially in actual fact.

Chapter 8: What Makes Quantum Computers Difficult To Make?

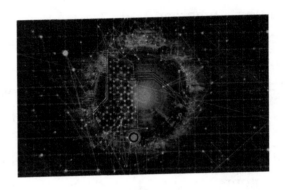

If they are not properly managed Quantum computers are significantly more error-prone than conventional computers. This extreme sensitiveness is due to the difficulty of keeping all qubits within the right superposition state for the entire process. For instance, it is important to remember that the states of the qubit superposition $(0)+(1)$ $((0))$ and $(0)+(-1)$ are distinct states. They both have the same probability for either a 0 or 1 outcome when the measurement is carried out however they are very different. Find a qubit that is described by only one photon to prove this. The state of polarization (H) + (V) is an inversely the polarized (D) condition, whereas (H) + ((-) is an oppositely Polarized (A) state. The state arrows of both states are perpendicular, which makes them totally different. These countries are extremely delicate. For instance, the accidental inclusion of a small timing variation between the V and H elements of the state diagonal could make it an antidiagonal. This could completely alter the quantum computation that was intended.

Computer errors caused by disturbances that are not expected or noise could prevent the quantum computer from giving correct answers in the event that there was no way to anticipate, detect the errors, and then correct them while the process is going. With the help of quantum theory, scientists have found ways to fix these errors in a functioning quantum processor. It is the idea to add additional qubits to the input, the aim being to track any errors that are not wanted. The extra qubits are interconnected in a unique, well-known way to the qubits we are interested in -the ones that do the computation. By measuring additional qubits without affecting the qubits we are concerned about, we can identify the possibility of an error having occurred. The error detection process is reminiscent however, not exactly with the detection of errors within Quantum encryption keys distribution system. If an error is discovered, it is rectified before proceeding with the calculation. Unfortunately, the addition of more qubits, that also require to be controlled flawlessly, adds a large quantity of complexity quantum computers which makes it extremely difficult to construct these computers.

What are the prospects of Building Quantum Computers?

It is a very difficult to answer question since the objective is a moving object. It is reasonable to conclude that there isn't a quantum computer that is universally operating. Numerous small-scale demonstrations have been conducted successfully and have shown that the physics upon which the potential of quantum computing is very strong. These tests seem to prove that building a quantum computer is currently just a matter creativity and a challenging engineering and not something that is fundamentally physics-related. Scalable implies that if one are able to build a quantum computer with, for instance 100 qubits then it's only twice as difficult to construct one that has two hundred qubits. It would be only three times as difficult to create one with three hundred qubits and the list goes on. If you could create a quantum processor which includes, say one hundred qubits it will be twice as difficult to construct one that has 200 qubits. This means that you don't want the effort in building it or the size and the resources required to be increased exponentially in proportion to the magnitude of the issue that you're trying compute. This could defeat the entire purpose of developing quantum computers to address the exponential scaling issue.

What are the best techniques to build Quantum Computers?

Although many techniques are being investigated but the three most promising technologies to build quantum computers include the superconducting circuits that are isolated electrons trapped magnetically in the vacuum chamber, or isolated Phosphorus atoms that are that are embedded within silicon crystals. However, the race to achieve this first, is worth an entire book. A nucleus that is composed of neutrons and protons forms the center of an atom of phosphorus. The nucleus acts as an extremely small bar magnet that has the north and south poles. The magnet can be placed with a north pole that faces up or down. The qubit can be represented as the magnet's orientation that is: UP is 1, and the opposite is zero. The direction of the nucleus magnet could be controlled for a brief period using a magnetic force the force generated by it causes the magnet to spin in an opposite direction. This allows for an quantum QR gate operation that is component of quantum computing process mentioned above. In order to build quantum computers, many phosphorus atoms must represent a variety of qubits. QXOR gate operations that require the qubits of two pairs need to be executed. The phosphorus atoms that make up the quantum computer are laid out in a pattern as a chessboard. There is one atom of phosphorus at the center of each square with dimensions of 30 nanometers in size by thirty nanometers. Keep in mind that one atom is approximately 0.2 nanometer, meaning that these are extremely small squares! In normal circumstances, the qubits in the magnet's inner axis of each phosphorus atom don't interfere with one another. This is known as 'quiet time' where the qubit values are just stored. The entire crystal of silicone must be kept at a low temperature, which is 391°F or the temperature of 196 degrees Celsius. This stops those internal magnetic elements from getting affected by the exaggerated movement of the silicone atoms that comprise the crystal. This can result in the nucleus's magnets being rotated accidentally in the opposite direction. We have already discussed this it could cause mistakes in the state of the qubit and require the application of error correction techniques. If researchers wish to run the QXOR gate in between two adjacent Phosphorus atoms, they will activate each one by passing an electron through nearby wires to the atoms. Because of the nuclear physics, the inner magnets in each atom are more powerful, and the two magnets tend to interact with one another. (If you've ever held two magnets in close proximity to one another, you'll know that the stronger magnet tends to push and turn the weaker one.)

The arrangement of two phosphorus atoms result in the actions of QXOR according to this The magnet in one of the molecules, known as qubit B, controls exactly what is happening to the magnet inside another atom of phosphorus known as qubit A. If B is equal to the value 0 (DOWN magnet) the qubit A is unaltered. If B is 1 (Magnet up) The magnet qubit A is pulled and rotated from 0 to 1 (DOWN to up) as well as from one to zero (UP to down).

Future Directions in Quantum Science

Quantum scientists must comprehend the complex nature of quantum phenomenon as well as their explanations using quantum theory to continue their progress. They must develop new methods for creating devices that are dependent on quantum phenomenon to operate. Quantum physicists work to better understand the non-local interactions that occur when measuring quantum-entangled particles.

Quantum theory is a method to explain the observed non-local correlations by using the concept of entangled states two objects. This kind of entanglement may occur even if the two entities are separated only by a large distance. The quantum explanation makes clear that the state of one particle doesn't directly impact the condition of the other particle, however, the correlations continue to occur in a way that is contrary to the classical description of Physics. Physics researchers would like to know the length of time these interactions last. While quantum theory describes these correlations precisely but it is not clear the mechanism by which they occur. Do you know if there is a backdoor channel that somehow transmits connections without violating the cause and causal nature of phenomena as is what the Theory of Relativity seems to need?

How do non-local correlations work when pair of quantum particles are in close proximity or inside a black-hole in which space-time is severely distorting? In such environments where space and time require us to consider what we are referring to when we refer to "local.' Such questions could provide discoveries in the understanding of the early and fundamental nature of the universe as a whole. Researchers in quantum technology are trying on increasing their expertise in the development and development of the devices required to improve the three major domains of study and development including quantum communications as well as quantum sensors and quantum computing.

Technology advancements are necessary for instance within the areas of: first is light sources that reliably produce single photons with predictable, controlled times. The problem is how to eliminate the random nature of quantum results in the creation of these photons. Another important breakthrough is tiny portable nuclear interferometers that can be used to detect inertial signals. These devices rely on the capability to cool a isolated cloud of atoms until temperatures close to absolute zero. Thirdly, strategies for building quantum computers that can scale--that is, a system that doubles the size of processors and memory only requires two times the amount of space and expense (not an exponential growth within memory). One example is the need to develop better techniques to place the atoms one at a time in a solid like silicon as well as a magnet trap within an evacuated chamber as well as methods to manipulate and verify the quantum state of their particles. These and many more are the subject of intensive continuing research. We aren't sure how far quantum technology could be developed to provide alternatives to conventional physics-based technology with greater capabilities. Can efficient, functional quantum computers be developed successfully? And If so, which jobs will they use most effectively? Will quantum-based sensors become more advanced and be utilized across a broad range of applications? What if the inherent nature of these instruments as well as their high sensitivity even minor interruptions make them ineffective?

The new technologies aren't likely to replace the existing ones and will instead complement them, and each is going to be employed to the extent that it is most appropriate. This is in line with the notion that the classical physics theory and quantum theory share a position and role for explaining the physical system. We choose the theory which is most suitable to address the issue in time. It is crucial to be aware of what you don't have, in order to advance in the field of science. It is crucial to ask the appropriate questions.

There are times when you may be confused by certain elements of quantum phenomenon, and their theories.

Most importantly, we've observed that the existence of things is probabilistic. That is, certain events happen in a largely random way. For instance the moment an electron gets exuberant to be in an energy state in an atom is likely to decay to lower energy, and release an electron. It can be destroyed at any point after the excitement has ended and there is no way of knowing precisely when this happens. However, causality remains in place and a particular event is not able to be triggered unless the required previous events have occurred and the prerequisite previous conditions are met. Based on certain events or conditions that have occurred prior to there exists many possible futures that each has the possibility of occurring at a particular time However, all of them must be compatible with the concepts of causality and impact.

There is currently a quantum theory that encompasses almost everything we know about quantum and classical. We don't have a complete understanding of the quantum theory's efforts to impart about Nature or at the very least not one that's shared by the overwhelming majority of scientists. There exists a group of physicists who are trying to come to understand the fundamental motives behind the phenomena we observe are best described using quantum theory and the structure it is able to take in terms of superposition, state arrows units, unitary processes, and other such. While they agree that the theory that we have today works, it also allows the creation and development of new quantum technologies scientists hope that studying the fundamentals of quantum theory on a basic, even at a philosophical level , will result in further discoveries.

Chapter 9: An Uncertain Heisenberg

And here's the opportunity that you've wanted to see for a long time. We're about to tackle the quantum mechanics of reality head-on and be thrown into an unknown and awe-inspiring terrain. The latest science has pushed to the point of frustration Wolfgang Pauli, one of the most influential physicists of all time that in 1925, in an email to a colleague stated that he was willing to end the battle: "Physics is now too complicated. I'd prefer to be a comic actor or something similar instead of being a physicist". If this titan of scientific thought had retreated from research in order to be like the Jerry Lewis of his time and we are now be discussing"the "Pauli exclusion rule" and the development of science might take a completely different turn.1 It was fortunate that Pauli persevered in the same way we hope you do in the following pages. The path we're going to take is not for those who are hesitant however, reaching the goal is an incredible reward.

Nature is created by packaging

We'll start with the quantum theory of old that was developed by Bohr to explain the findings that resulted from the Rutherford experiment. As you may recall, it was a replacement for the "panettone" model of the atom, with the notion it was the existence of a large central nucleus, surrounded by electrons that whizzed around - an arrangement similar to that of the solar system and our star in the center, and the planets moving around it. We've already mentioned that the model also passed to a more favourable state. As a result of further refinements the original quantum theory that was a wacky mix of classical mechanics and ad-hoc quantum modifications, at one point abandoned completely. The benefit of Bohr was that he presented an unprecedented opportunity to present an atom quantum model that gained credibility due to the results of his remarkable experiment that we'll soon be witnessing.

Based on the classical laws that no electron can ever be able to remain orbiting around the nucleus. The motion of the electron would be increased in the same way as any circular motion (because the speed is constantly changing direction over time) as per Maxwell's laws, every charged particle moving at a high speed emits energy through electromagnetic radiation, also known as light. Based on classical calculations that an electron orbiting is likely to lose all of its energy and disappear as electromagnetic radiation. Therefore, the particle concerned lost altitude and will shortly crash into the nucleus. The classic atom cannot exist, if it wasn't in collapsed form, which makes it chemically dead and not serviceable. The traditional theory was not able justify the energy levels of nuclei and electrons. Therefore, it was necessary to develop a brand new model, the quantum theory.

Furthermore, as was well-known at about the time of the 1800s due by the spectral line, electrons emit light, but only in specific colors, that is, they emit frequencies (or frequencies) that have quantized, discrete values. It appears that there are only certain orbits that exist, and electrons bound to move from one orbit to the next whenever they emit the energy or take in. If this were the case, then according to it was the "Keplerian" theory of the atom as a solar system and the spectrum of the emitted radiation would be continuous because classical mechanics permit an infinite spectrum of orbits. Instead, it appears that the universe of the atom is "discrete" and is very distinct from the continuity that is provided by Newtonian Physics.

Bohr concentrated upon the most basic Atom of all that is the hydrogen one with just one proton in the nucleus, and an electron that orbits around it. While playing around with the latest ideas in quantum mechanics he realised that electrons could be subjected to the Planck theory, which is to link to a particular frequency (or frequency) the energy (or the energy) that a light particle has from which it can be determined the existence of discrete orbits. After several attempts, he arrived at the correct formula. It was discovered that Bohr's "special" orbits are circular, and each was assigned a circumference always at a level that was equal to quantum wavelength electrons derived from the Planck equation. Magic orbits were associated with specific energy levels, so the atom was one distinct set of states in energy.

Bohr immediately recognized the existence of an orbit that was a minimum, in which the electron was near as is possible close to its nucleus. From that point, it was impossible to fall lower, so the atom could not be expected to fall apart and thus its final fate. This orbit is referred to as the fundamental state. It is the lowest energy state for the electron. The existence of this state indicates that the stability and integrity of an electron. We now know that this property is a characteristic of every quantum system.

Bohr's theory was very effective in that, from the new equations, all numbers that matched the values found in the experiments came out one after the next. The electrons in the atom are would say, "bound" and without the energy source from outside, they continue to rotate about the nucleus. It is the amount required for them to blast away and release them from their atomic bonds is known as bonding energy. It is determined by the orbit on where the electron is situated. (Usually we refer to this as the minimum amount required to push the electron from an atom , and then bring it to an the limitless distance, and without energy in kinetics, which is what we usually consider to be zero energy, however this is simply a convention). In the reverse direction, when an electron that is free is trapped by an atom it releases energy by way of photons, that is equal to the bonding on the orbit it is in when it comes in.

The bonding energies in orbits (i.e. states) are typically measured in units known as electronvolt (symbol called the symbol eV). The hydrogen fundamental state is atom, which corresponds to the special orbit with minimum distance from the nucleus and highest bonding energy, is equivalent to 13,6 eV. This number can be calculated by using the"Rydberg's formula" or the "Rydberg's formula" that is named in memory of the Swedish scientist Johannes Rydberg, who in 1888 (extending certain research conducted by Johann Balmer and others) provided an explanation based on empirical research of the lines in the spectral spectrum that exist in hydrogen as well as other elements. Actually, the number 13,6 electrons, and how it could be calculated were well-known for a few years, however Bohr is the one to provide an elucidative theoretical explanation.

The quantum state of electrons in the hydrogen atoms (equivalent with one Bohr's orbits) are represented by an integer number of n = 1,2, 3, ... The state with the greatest bonding energy, called the fundamental one, is one; the initial excited state is n=2, and the list goes on. It is the fact that this distinct state set is the only possible one in atoms is at the heart the quantum theory. The number n enjoys the distinction of having being given its own name in physics. It is known as "main quantum numbers". Every quantum state quantum number is distinguished by an energy value (in eV, similar to the one earlier) and is identified by the alphabets E1, E2, E3 and so on.. (see the note at 3).

You will recall that in this model, old but not ignored, it is believed that electrons emit photons which jump from an energy higher state to the lower energy state. This does not apply to the state of fundamental E1, which occurs when the n=1 value is present because in this situation, the electron does not have a lower orbit that is available. The transitions occur in a totally rational and predictable manner. For instance, if when the electron of the state n=3 moves to the state of that n=2, the person who is in the last orbit needs to go down to the state n=1. Every jump is followed with the release of a light particle with an intensity equal to that of the differences in the energy levels of the concerned states, for instance E2-E3 or E1E2. In the case of hydrogen atom that is the case, the numerical values are 10,5 eV - 9,2 EV = 1,3 eV and 13,6 eV minus 10,5 eV = 3,1 EV. Because energy E and the wavelength of a photon is connected by the formula of Planck E=hf=hc/l. It is possible to determine the energy of photons emitted by measuring their wavelengths thanks to spectroscopy. The

Bohr's times were when the accounts were resurfacing on hydrogen atoms. It was the most basic (only one electron within proton) however, since it was already the vicinity of helium which is the second element, in order in terms of its simplicity didn't know the best way to proceed.

In Bohr came up with a different idea which was to determine the momentum of electrons through absorption of energy by electrons in the atoms, which reverses the logic that was previously discussed. If the theory of quantum states is correct and atoms absorb energy in only packets that correspond to the different states' energy levels, E2 versus E3 or E1 - E2 and the list goes on. The key experiment to test of this hypothesis was carried out during 1914 by James Franck and Gustav Hertz in Berlin and was the final significant research done in Germany prior to the start in the First World War. The two scientists came up with results that were in complete agreement with Bohr's theory that they did not know about. They would have been aware of the findings of the famous Danish physicist just a few years later.

The terrible twenty-first century

It's hard to grasp the fear that was rife among the top physicists in the world at the start of the horrific 1920s, from 1920 until 1925. Following four hundred years of believing on the rational principles that underlie the nature of things science was suddenly being forced to review its own base. One of the things that most shook minds, while being lulled by affirmative tenets from the past was the confusing quantum theory's dual nature. On one side, there was ample experimental evidence to show that light behaved as waves, with lots of diffraction and interference. As we've seen in depth the wave hypothesis is the only theory that has the ability to explain the results that were gathered from the double slit experiment.

However there was a similarly large amount of observations strongly demonstrated the nature of light as a particle as we witnessed it in the previous experiment with black body radiation Photoelectric effect, black body radiation as well as the Compton effect. The conclusion that the experiments led to was a single and unambiguous one the fact that light of any hue and therefore that of all wavelengths, is formed by a stream of particles that were all moving through the vacuum at the same rate. Each particle was a different momentum, which in Newtonian Physics was calculated by the mass product for speed. The same quantity for photons is that energy multiplied by the number c. The importance of momentum (as can be demonstrated by any person who walked through an speed camera in a car) since its total within a system is maintained and doesn't change after multiple collisions. In the case of classic collisions, it is the story of two balls colliding. Even if speed changes the sum of the momentum prior to and following the collision is constant. Compton's research has demonstrated the conservation of momentum is applicable to the various forms of light molecules, that in this way behaves like automobiles and other macro-sized objects.

It is best to take one moment to explain the distinction between particles and waves. First , the second kind of particles are distinct. You can use two glasses that are filled with water and the other with fine sand. Both of them change in shape and are able to be poured in a manner that upon a less careful examination, they appear to have the same characteristics. However, the liquid is constant and smooth. The sand is comprised of small grains that are accounting. While a teaspoon has some amount of liquid. A teaspoon of sand is quantifiable by the amount of grains. Quantum mechanics is a method of evaluating numbers and discrete quantities, as if like a returning towards Pythagorean theories. The particle, in each moment, is in a predetermined location and follows the same path in contrast to a wave that is "smeared" across space. Furthermore, particles have a specific energy and momentumthat can be transferred to other particles during collisions. In the definition of a particle, it is not a wave, and reverse.

Then we come back. Scientists in the 20th century were baffled by the strange beast, which was semi-wave and particle known as wavicle by some. (wave or wave and particle, contraction of particle). In spite of the evidence that has been accumulated that supports the idea of wave nature in experiment after experiment, photons were discovered to be real objects capable of colliding with one the other as well as with electrons. Atoms emit one each time they were in an excited state, emitting the identical amount of energy, E=hf, which was emitted by the photon. The story took an even more surprising turn with the entry of a young French physicist, the aristocrat Louis-Cesar-Victor-Maurice de Broglie, and his memorable doctoral thesis.

The de Broglies, with a small number of the highest-ranking diplomats, officers as well as politicians, was opposed to Louis his inclinations. The duke of his time, his father, called science "an old woman seeking the attention by young males. The scion, in the desire of compromise, enrolled in studying to make a naval officer however, he continued to study in his spare time because of the lab he set up at the ancestral home. While in the navy, was where he established himself as a specialist in broadcasting. After the death of the previous duke, he was allowed to leave in order to devote his entire time to his passion.

De Broglie had thought at long length about Einstein's doubts regarding this photoelectric phenomenon, which seemed inconsistent with the nature of light's undulating turbulence and that confirmed the hypothesis of the photon. When rereading the work of Einstein's renowned scientist his young Frenchman came with an unusual concept. When light that may appear to be a wave, has a particle-like behaviour and in the natural world, then perhaps there is a reverse. Perhaps particles, or any particle, exhibit certain situations, behaviors that are similar to wave. De Broglie's words "[Bohr the theory of atoms] led me to believe that electrons might be thought of as not just particles, but things to which it is capable of assigning a frequency which is a wave characteristic".

In the past the year 1924, a doctoral student who chose this risky idea for his dissertation could have been forced to the Theology faculty of a secluded university in Molvania Citeriore. However, in 1924, everything was feasible in addition, de Broglie had a very unique admirer. The legendary Albert Einstein was called by his confused Parisian colleagues to be an external consultant to study the thesis of the candidate and found it very fascinating (perhaps the thought was "but what if I had this thought before?" ?"). In his letter for the Parisian commission, Einstein said: "De Broglie has lifted the veil of great importance". This teenager Frenchman not only won this title but a couple of years later, he was given the Nobel Prize for physics, due to the concept he developed within his dissertation. His biggest achievement was discovered a connection, modeled after Planck's, the classic acceleration that an electron has (mass in reference to its the speed) as well as the frequency of its wave. But what exactly is the wave? Electrons are particles For goodness sake What is the location of the wave? De Broglie was talking about "a mysterious phenomenon that has periodicity-related characters" that occurred inside the particle. It's not clear but he was certain that it was true. Even though his interpretation was smokey however, the concept behind it was truly brilliant.

In 1927, two American physicists , working at the famous Bell Laboratories of AT&T, New Jersey, were studying the properties of vacuum tubes that bombarded with electron flows of various kinds of crystals. The results were rather bizarre that the electrons were flowing out of crystals in the direction they preferred and were unable to discern other directions. They Bell Lab researchers were not in a position to achieve this until they learned about the baffling hypothesis of de Broglie. In this new viewpoint it is clear that their experiment was an advanced version of Young's featuring a double slit and the way in which electrons demonstrated a known property of waves, which is diffracted light! The findings would have been logical in the event that the length of electrons was in fact dependent on their speed, exactly as de Broglie predicted. The lattice that is regular to the crystals' atoms is the same as the cracks observed in Young's experiment which occurred nearly a century prior to. This discovery that was fundamental to "electronic difffraction" confirmed de Broglie's theory electrons are particles that behave as waves, and it's also easy to confirm that.

We'll be back within a short time on Diffraction, and re-creating the familiar double slit test using electrons. This will yield a more baffling result. We can only see that this phenomenon is responsible for the fact that different materials behave like insulators, conductors, or semiconductors. This is the reason for transistors' invention. We now have to meet a new persona - possibly the real quantum revolution's superhero.

A strange mathematical concept

Werner Heisenberg (1901-1976) was the godfather of the theorists. He was who was so uninterested in the laboratory that he feared failing his dissertation in the University of Munich because he was unsure of how batteries functioned. Fortunately for him and for physics in general that he was promoted. There were also some difficult events during his lifetime. Through during the First World War, while his father was on the front fighting as soldier, the shortage of fuel and food for the capital city of Berlin was so severe that universities and schools were often forced to cancel classes. In the summer of 1918, young Werner who was weak and hungry and hungry, was compelled along with his fellow students to assist farmers in an Bavarian harvest on the farms.

After the war, during the beginning of the 20th century, we see him as the young star: a musician of the highest level with a strong background in classical languages, a skilled skier and alpinist and a mathematician of high standing, affixed to the Physics. While studying with his older teacher Arnold Sommerfeld, he met another young talent, Wolfgang Pauli, who was to become his close friend and fiercest critic. In 1922, Sommerfeld took young 21 year old Heisenberg to Gottingen which was then the center for European scientific research, in order to see the series of lectures that were devoted to the emerging quantum atomic physics. The lectures were given by Niels Bohr himself. At that time, the young researcher, who was not too frightened, attempted to defy the claims of the guru, and even challenge at the core of his theoretical model. After the first encounter there was established a long-lasting and successful collaboration that was that was marked by mutual admiration.

From that point on, Heisenberg committed his both body and spirit to mysteries that are quantum mechanics. In 1924, he stayed for a short time in Copenhagen to collaborate directly with Bohr on the issues of radiation absorption and emission. In Copenhagen, he was able that he could appreciate the "philosophical approach" (in Pauli's terms) of the legendary Danish scientist. Incredulous of the difficulty to translate the model of the atomic structure of Bohr and its orbits placed in such a manner that is not clear the young man became convinced that something was wrong at the core. The more he considered it the more it appeared to him that the simple circular orbits were just a waste and purely an intellectual concept. To eliminate these orbits, he began to think that the idea of orbit was an Newtonian leftover that needed to go.

The young Werner set himself up with a strict principle: every model was to be based on the classical physical science (so no mini solar systems, even though they're so adorable for drawing). The only way to get back was not a matter of aesthetics or intuition and mathematical precision. One of his dikats conceptually was to renounce any entity (such like orbits in actual) that were unable to be directly measured.

In the atoms, measurable could be seen in the spectrum lines. They were evidence of the absorption or emission of light particles by particles as a result of shifting between electron levels. It was therefore to these visible and verified lines, which correspond to the inaccessible subatomic world which Heisenberg was able to direct his focus. To resolve this diabolically complex issue, and also to get relief from hay fever, in the year 1925 Heisenberg retreated to Helgoland an isolated island in the North Sea.

The basis of his work was the known "correspondence principle" that was articulated by Bohr who stated that quantum laws must be converted without difficulty into classic laws when applied massive systems. How big? enough to be able to ignore to consider the Planck constant of h within the equations of relative. The typical object in the atomic universe has a mass of 10 to 27 kg. Let's take for instance that a particle of dust not visible for the naked eye may weigh 10-7 kilograms, which is incredibly tiny, yet nevertheless larger by the factor 100000000000000. That's 1020. This is followed by 20 zeros. Therefore, the atmospheric dust is in the realm of classical physics. It is a macroscopic object , and its motion isn't dependent on the existence of any factors that are dependent on the Planck constant. The fundamental quantum laws are applicable naturally to the phenomenon of the subatomic and atomic realms however it becomes unwise using them to explain things that occur in aggregates greater than atoms, when the dimensions expand and quantum physics is relegated to the traditional principles from Newton as well as Maxwell. The basis of this concept (as we'll repeat it repeatedly) lies in the bizarre quantum effects that are not published "correspond" in direct correspondence to fundamental concepts of physics when you depart from the atomic field and enter the macroscopic field.

In the spirit of Bohr's theories, Heisenberg changed in quantum field some of the most common concepts of classical physics such like the velocity and position of an electron, in order it was in harmony with Newtonian equivalents. He soon realized his efforts to reconcile two worlds lead to the development of a strange and new "algebra of physical physics".

We were all taught in school the commutative property, also known as multiplication. This is the truth that, in the case of two numbers, a and b, their product will not change when we swap the two numbers; for example, in the form of symbols like $A \times b = b \times a$. It is evident that, for instance, $3 \times 4 = 4 \times 3 = 12$. However, in Heisenberg's time it was evident that there were abstract numerical systems, in which the commutative property doesn't always hold and there is no way to say that the x b of a is equivalent to bx A. When you consider it the examples of non-commutative operations can be found in the natural world. One of the most common examples is tilts and rotations (try to make two distinct rotations on an object , such as book and you'll see instances where the sequence that they occur is crucial).

Heisenberg did not study deep into the most advanced areas of pure mathematics of the day, but He was able to enlist the assistance of more knowledgeable colleagues who quickly realized the kind of algebra in his definitions. They were nothing more than the multiplication of matrices having complicated values. The term "matrix algebra" was a fanciful mathematical branch that was in use for around sixty years. It was used to describe objects made up of columns and rows of numbers as the matrices. Matrix algebra, when applied to Heisenberg's formalization (called"matrix mechanics") resulted in the first abstract arrangement of quantum Physics. His calculations led to useful results for the energy of states as well as atomic transitions which are electron level leaps.

When the matrix mechanics concept was applied not just to the hydrogen atom instance as well as to other microscopically-based systems, we found that it performed flawlessly The solutions that were found theoretically matched the experimental evidence. From these baffling matrix manipulations, there emerged as a new concept.

The very first step of the principle of uncertainty

The primary result of non-commutationality was revealed as this. If we show by x the location on an axis and p the speed, all in the same direction, of particles however, the fact that xp isn't equal to px indicates that the two numbers cannot be measured in a specific and precise manner. Also that if we determine the exact location of a particle, we break the system in that it's impossible to determine the momentum of that particle as well as the reverse is true. The reason for this isn't technological and it's not the instruments we use that are incorrect Nature is the reason why it has been designed this way.

In the formalization of matrix mechanics, we can put this concept in a succinct manner that has been a source of confusion for philosophers of science. mad: "The uncertainty relative to the location of particles, as indicated by Dx and also that in relation to the movement, Dp, are linked through the equation to: DxDp>=h/2, in which H=H/2p". In other words "the result of the uncertainties related to the location and velocity of a particle is constant or greater to an amount that is equal to Planck constant multiplied 4 times". This means that if you determine the position with extreme precision which makes Dx as small as is possible then we automatically make Dp in a way that is excessively large and the reverse is true. It's impossible to have everything and we must stop knowing the exact details of the location, or the acceleration.

Based on this concept, we can also determine the stability of the Bohr Atom, which is to show the existence of an essential state, an orbit that is lower, under which electrons cannot descend like it does within Newtonian mechanics. If the electron moves closer and closer towards the nucleus until it reaches with us, uncertainty regarding its position would become less and less the reason scientists say Dx "would decrease to zero". According to the Heisenberg principle Dp will become arbitrarily high which means that the electron's energy would increase more. It is demonstrated that there exists an equilibrium condition in which the electron has "enough" and well-placed, with Dx differing from zero and the energy is the smallest achievable, given the Dp value.

It is much easier to comprehend if we place ourselves in a different order of reasoning, similar to Schrodinger and then look at the property (not quanta) in electromagnetic radiation which is well-known in the field of telecommunication. Yes, we're returning to the wave. The matrix mechanics appeared initially to be the only method that could be used to get into the tangles of the atomic universe. However, fortunately, just as scientists were about to become experts in algebra In 1926, another appealing solution was offered.

The most beautiful mathematical equation in all of history

We already met Erwin Schrodinger in chapter 1. You may recall that at some point, he went on his vacation in Switzerland to do his research in peace.

The outcome of his time was an equation, Schrodinger's equation that brought a lot of understanding to quantum phenomena.

What is the significance of it? Let's revisit Newton's first law, F=ma law that regulates the motion of planets, apples, as well as all macroscopic objects. It says that applying force F to an object that has mass m causes the acceleration (i.e. the change in velocity) to and from are connected by the equation described above. The solution to this equation lets us know the current state of a particular body, such as tennis balls in every single moment. The most important thing to do is to determine F, from which the position x as well as V's velocity at time t occur are calculated. The relations between these numbers are established using differential equations that employ theories that use infinitesimal analyses (invented by Newton himself) and can be complicated to answer (for instance, when the system is comprised of several bodies). The form of these equations however is relatively simple. They are calculations, and they allow their applications can become complex.

Newton impressed the world with his discovery the connection between the universal law of gravitation with that of motion, when applied to gravity's force and the simple elliptical orbits as well as planetary motion laws which Kepler had articulated in the Solar System were discovered. The same equation can be used to explain the motion of the Moon and that of an apple dropping off the tree, and of an orbiting rocket. The equation cannot be solved in a clear manner if more than four bodies are involved, and all of which are subject to gravitational influence In this situation, it is required to solve through approximations, or with the aid of numerical techniques (thanks for calculators). This is a great example where at the heart of the natural laws there's an incredibly straightforward formula, but it's a testament to the staggering magnitude of our universe. Schrodinger's equation can be described as the quantum form of F=ma. If we solve it but we do not get the position and velocity of the particles as in the Newtonian situation.

The holiday in December 1925, Schrodinger took with him not only his wife and an original copy of de Broglie's doctoral dissertation. A few people, at the moment, were aware of the thoughts of the Frenchman however, after Schrodinger's studying the thesis, things changed drastically. In the month of March 1926, Schrodinger was a forty-year-old professor from Zurich's University of Zurich, who was, at that time, not having an enviable career, and who, according to what was considered to be a youthful physicists at the day was impervious, announced to the world at large his equation that addressed the motion of electrons as waves that was based on de Brglie's thesis. His colleagues thought he was more palatable than the stoic abstractions in matrix mechanics. In Schrodinger's equation, a new fundamental number was discovered as the wave function marked with the symbol Ps, symbolizes its solution.

Before the introduction of quantum mechanics physicists have been used to study (classically) diverse types of material waves that are continuous space, including sound waves that travel through the air. Let's take a look at an example involving sound. The thing that we are interested in is the pressure created by the sound wave in the air, which we represent by using Ps(x,t). From a mathematical point perspective, this is an example of a "function" that is a formula which calculates the pressure of the wave (intended to be a variation from normal atmospheric pressure) at every point x in space, and at every moment at each time point. The solutions to the equation of the relative classical type naturally define a wave that "travels" through both time and space "perturbing" movement of particles in air (or water or electromagnetic fields or any other). Waves of ocean or tsunamis, as well as beautiful people are all examples of the forms that can be derived from these equations and can be classified as "differential" kind: they contain variables that fluctuate in a way that is understandable by it is necessary to understand the mathematical formula. "The "wave equation" is a form of differential equation which if solved, gives an answer known as"the "wave equation" $Ps(x,t)$ which is in our case, the pressure of air which changes in time and space as the sound waves pass through.

With the help of de Broglie's concepts, Schrodinger immediately understood that Heisenberg's intricate technicalities could be modified to produce relationships that are similar to the classic mathematical equations in classical physics. particularly the equations for waves. From a formal point perspective, a quantum particle was represented by the function Ps(x,t) which Schrodinger himself described as a "wave function". Based on this interpretation and applying the fundamentals of quantum physics i.e. solving Schrodinger's equation it was feasible in principle to determine the wave function of every particle that was then identified in the majority of cases. However, nobody had any idea of what the value of this number was.

Because from the invention of Ps we are no longer able to claim that "at time t that particle exists in the x" but instead, we have to state"that "the movement of the particle can be described through the formula Ps(x,t) which is the magnitude Ps at that t is at x". The exact location is not determined. If we notice that Ps is especially massive at a particular point and is almost non-existent elsewhere it is possible to say it is "about in the position in the x position". Waves are a collection of objects in space, as is the function of the wave. We see that these are logical reasonings with hindsight as in the period we're examining, nobody even Schrodinger had any specific ideas about the essence of the function of waves.

There is, however, an interesting twist that is an extremely intriguing features that quantum mechanics has to offer. Schrodinger discovered the wave functions of his were like one would expect from a wave constant in time and space however, to bring the numbers back, he needed to take various numbers as values that differed from the actual ones. This is a huge distinction from regular waves, whether electromagnetic or mechanical, whose they are always in the real. For instance, we could claim that the crest of one of the waves in oceans rises above the sea's average level by 2 meters (and this is why we need to display the red flags on the beach) and even more alarmingly that a tsunami of 10 meters high is on the way and that is why we have to evacuate coastal zones in a short time.

These are real numbers that are concrete and quantifiable using different tools, and we recognize their significance.

Quantum wave functions however it is an assumption of values within the realm of "complex numbers".15 For instance it could be that at x, the amplitude is equivalent to the value of a "stuff" which is written as 0.3+0.5i (i=1). That is multiplying the number i by itself yields the result of -1. An object similar to the one described above, which is real numbers that is added to a actual number, multiplied using the number i, is considered to be the term "complex number. Schrodinger's equations always require the presence of the number i. It is a key element to the calculation itself. This is the reason why the wave function takes on complex values.16

This mathematical problem is an inevitable step in the direction of quantum physics. It's also an proof that the function of waves of a particle isn't easily quantifiable:

after all real numbers are only found during tests. In Schrodinger's perspective the electron is a wave that can be used for every purpose, and is not distinct from a sound or a marine wave. But how can this be possible that the particle must be placed at a specific point and is unable to occupy large areas of space? The trick is to superimpose a variety of waves in a manner that they disappear almost everywhere, except for the area we're looking for. A mixture of waves could therefore, represent an object that is well-placed in space. We might be tempted to label "particle" and pops out every when the sum of the waves results in an a specific concentration at the form of a point. In this sense , a particle could be considered to be an "anomalous wave" like the phenomena that is caused by the interplay of waves that causes a massive disturbance that can cause boats to turn.

An eternal adolescent

When did quantum theory at the point of discovery following the discoveries from Heisenberg, Schrodinger, Bohr, Born and colleagues? There are probabilistic wave functions on one hand, and the principle of uncertainty on the other hand, that allows for the model of particle. The problem of duality "a bit wave and a particle" appears to be resolved as photons and electrons are particles, and their behaviour is described using probabilistic waves. Because they are waves, they are susceptible to interference effects which cause the docile particles at the exact location they are supposed to appear, based on waves. The way they do it isn't a concern which is reasonable to discuss. They say this in Copenhagen. The cost of success is the intrusion into the physics of probability as well as a myriad of quantum anomalies.

The notion that the universe (or God) plays dice by playing with subatomic matter did not attract Einstein,

Schrodinger, de Broglie, Planck and many others. Einstein especially believed of the fact that quantum mechanics was just an unfinished stage, a temporary theory that would soon and later replace with an alternative, causal and deterministic one. In the second half of his life, the great scientist attempted several times to solve the issue of uncertainty However, his attempts were rejected every time by Bohr which was supposedly to his naive satisfaction.

Therefore, we must conclude the chapter that was snagged between the successes of the theories and a sense of unease. In the 20th century, quantum mechanics had become an adult science but still prone to change and change. It was modified several times until the late forties.

Conclusion

The most important thing to be aware of regarding Quantum Physics is that when electrons aren't being observed, they behave as waves that display wave-like patterns. When electrons are watched and manipulated, they transform into particles of matter and exhibit what patterns appear on a single line one would imagine to shoot particles through openings. What exactly does this mean? Let's see how we can take it. Prepare yourself to...

There's nothing distinct. The entire universe including us are energy in pure form. Through our thinking that we transform this energy into something we consider reality. Remember, in the film the waves changed into particles after being seen. Our thoughts create the environment in which we reside. The fundamental shift is that we've always believed that the outer world is more accurate than the inner one. The reverse is the case of this. The inner workings of us is what determines what will occur out there. Through our thoughts we make our own world.

Ask yourself this the following question. What is it that means for you to have energy play guitar, sing or to speak French or install computers? In our 3D world our brains use our mind to perceive that energy is a physical manifestation. This is why we play the piano, the guitar and more. We are the force that governs everything throughout the world. This means it can't be a part of any one person. If you're able to do it with a single person then you can do it with any other guy.

When energy waves are observed, they break into particles. This is how our world is created. Our universe is constructed by crashing waves of possibility into our our mind. Probability is the method of calculating the likelihood of an event to occur. It is it is a number which represents the ratio of possible cases to the totality of events that could occur.

We have basically an endless amount of probabilistic wave functions that could actually fall apart. What you choose to break into reality is entirely your choice. There are literally many options among which could include guitar playing, piano singing, singing and speaking French or computer programming.

What we need to accomplish is to increase the chance of bringing things we'd like to see in our lives and reduce the chance of bringing things we do not want to see into our lives.

What are we going to do? accomplish this?

We transmit the vibrations of our thoughts or, to be more precise through the emotions our thoughts produce. Different emotions vibrate at different frequency. The positive emotions vibrate higher frequency. The negative emotions are at lower levels. In order to increase the chance of crashing this probability wavefunction, if you are looking to transform your dreams into reality, it is essential to be focused on what you would like to achieve and feel positive emotions.